THE PRINCIPLES OF AGRICULTURE

A TEXT-BOOK FOR
SCHOOLS AND RURAL SOCIETIES

EDITED BY

L. H. BAILEY

EIGHTH EDITION

British Library Cataloguing-in-Publication Data
A catalogue record for this book is available from
the British Library

Liberty Hyde Bailey

Liberty Hyde Bailey was born on 15 March 1858 in the small town of South Haven, Michigan, USA. He was the third son of farmers Liberty Hyde Bailey Sr. and Sarah Harrison Bailey and possessed a keen interest in horticulture and botany from an early age. Bailey entered *Michigan Agricultural College* in 1878 and graduated four years later. In 1883 he became assistant to the renowned botanist Asa Gray; a remarkable achievement for a young man straight out of university. The same year as this success, Bailey married Annette Smith, with whom he had two daughters, Sara May, born in 1887, and Ethel Zoe, born in 1889. Moving on from his apprenticeship with Asa Gray, Bailey moved to *Cornell University* in Ithaca, New York in 1885 and was appointed chair of Practical and Experimental Horticulture three years later. He enjoyed considerable success in this position and became an Associate Fellow of the *American Academy of Arts and Sciences* in 1900.

Bailey's incredible rise throughout the academic and horticultural world continued on his appointment, by Theodore Roosevelt, as Chairman of the *National Commission on Country Life* in 1908. Roosevelt was a renowned lover of America's farmland and countryside, and welcomed the 1909 report of the commission which called for rebuilding a great agricultural civilisation in America. Bailey strongly believed, in an agrarian tradition harking back to Thomas Jefferson, that rural civilisation was a vital and wholesome alternative to impersonal and corrupting city life. He especially endorsed family life, and the family farm as having a benign influence on societal responsibilities. Bailey's real legacy was the themes and direction he gave to the new agrarian movement however, promoting inclusive as opposed to exclusive sociability, as well as welcoming technological progress. Bailey retired in 1913 to become a private scholar and devote more of his time to social and political issues. Before this date though, he was very involved in editing academic works; *The Cyclopedia of American Agriculture* (1907-09) and the *Cyclopedia of*

American Horticulture (1900-02). He was also the founding editor of the journals *Country Life in America* and the *Cornell Countryman.* Bailey dominated the field of horticultural literature, and in total wrote sixty-five books, which together sold over a million copies. His most significant contributions to the field were in the botanical study of cultivated plants, notably emphasising the importance of Gregor Mendel's work on cross breeding and hybridizing. Bailey died on Christmas Day, 1954. He has been memorialised at *Cornell University*, by dedicating *Bailey Hall* in his honour as well as *Michigan State University* who created the Liberty Hyde Bailey Scholars program, designed to incorporate Bailey's love of learning with the wider expression and dissemination of this knowledge.

PREFACE

The greatest difficulty in the teaching of agriculture is to tell what agriculture is. To the scientist. agriculture has been largely an application of the teachings of agricultural chemistry; to the stockman, it is chiefly the raising of animals; to the horticulturist, it may be fruit-growing, flower-growing, or nursery business; and everyone, since the establishment of the agricultural colleges and experiment stations, is certain that it is a science. The fact is, however, that agriculture is pursued primarily for the gaining of a livelihood, not for the extension of knowledge: it is, therefore, a business, not a science. But at every point, a knowledge of science aids the business. It is on the science side that the experimenter is able to help the farmer. On the business side the farmer must rely upon himself; for the person who is not a good business man cannot be a good farmer, however much he may know of science. These statements are no disparagement of science, for, in these days, facts of science and scientific habits of thought are essential to the best farming; but they are intended to emphasize the fact that business method is the master, and that teachings of science are the helpmates.

But even if these facts are fully apprehended, the teacher and the farmer are apt to make no distinction between the fundamental and the incidental applications of science, or between principles and facts. Therefore, the mistake is often made of teaching how to overcome mere obstacles before explaining why the obstacles are obstacles. How to kill weeds is a mere incident; the great fact is that good farmers are not troubled with weeds. Rather than to know kinds of weeds, the farmer should know how to manage his land. How to know the weeds and how to kill them is what he calls practical knowledge, but, standing alone, it is really the most unpractical kind of knowledge, for it does not tell him how to prevent their recurrence year after year. The learner is apt to begin at the wrong end of his problem. This is well illustrated in the customary discussions of under-drainage. The pupil or the reader is first instructed in methods of laying drains. But drainage is not the unit. The real unit is texture and moisture of soils: plowing, draining, green-cropping are means of producing a given or desired result. The real subject-matter for first consideration, therefore, is amelioration of soil rather than laying of drains. When the farmer has learned how to prepare the land, and how to grow plants, and how to raise animals, then he may enquire about such incidental details as the kinds of weeds and insects, the brands of fertilizers, the varieties of apples, when he shall till, whether he shall raise wheat or sweet corn. The tailor first learns how to lay out his garment; but the farmer too often wants to sew on the buttons before he cuts his cloth.

Again, the purpose of education is often misunderstood by both teachers and farmers. Its purpose is to improve the farmer, not the farm. If the person is aroused, the farm is likely to be awakened. The happy farmer is a more successful farmer than the rich one. If the educated farmer raises no more wheat or cotton than the uneducated neighbor, his education is nevertheless worth the cost, for his mind is open to a thousand influences of which the other knows nothing. One's happiness depends less on bushels of corn than on entertaining thoughts.

Not only do we need to know what agriculture is, but we should know the relative importance of its parts. It is commonly assumed that fertilizing the land is the one most fundamental thing in agriculture, but this is not so; for if but one thing about farming practices were to be explained, that thing should be the tilling of the land.

Agriculture, then, stands upon business, but science is the staff. Business cannot be taught in a book like this; but some of the laws of science as applied to farm-management can be taught, and it is convenient to speak of these laws as the principles of agriculture. These principles are arranged in a more or less logical order, so that the teacher may have the skeleton of the subject before him. The subject should not be taught until it is analyzed, for analysis supplies the thread upon which the facts and practices may be strung. The best part of the book, therefore, is the table of contents.

A book like this should be used only by persons who know how to observe. The starting-point in the teaching of agriculture is nature-study,—the training of the power to actually see things and then to draw proper conclusions from them. Into this primary field the author hopes to enter; but the present need seems to be for a book of principles designed to aid those who know how to use their eyes.

L. H. BAILEY.

HORTICUTURAL DEPARTMENT,
CORNELL UNIVERSITY, Dec. 1, 1898.

ANALYSIS

INTRODUCTION (pages 1–15)

PART I

THE SOIL

CHAPTER I

THE CONTENTS OF THE SOIL (pages 16–36)

CHAPTER II

THE TEXTURE OF THE SOIL (pages 37–46)

(By JOHN W. SPENCER, Deputy Chief Bureau of Nature-Study in the Cornell University)

CHAPTER III

The Moisture in the Soil (pages 47–63)

(By L. A. Clinton, Assistant Agriculturist, Cornell University)

CHAPTER IV

The Tillage of the Soil (pages 64–76)

CHAPTER V

Enriching the Soil — Farm Resources (pages 77–86)

CHAPTER VI

Enriching the Soil — Commercial Resources (pages 87–105)

(By G. W. Cavanaugh, Assistant Chemist to the Cornell Experiment Station)

Part II

THE PLANT, AND CROPS

CHAPTER VII

The Offices of the Plant (pages 106–111)

CHAPTER VIII

How the Plant Lives (pages 112–131)

(By B. M. Duggar, Instructor in Botany, Cornell University)

CHAPTER IX

The Propagation of Plants (pages 132–144)

CHAPTER X

Preparation of Land for the Seed (pages 145–158)

(By I. P. Roberts, Director of the College of Agriculture, Cornell University)

CHAPTER XI

Subsequent Care of the Plant (pages 159–178)

CHAPTER XII

PASTURES, MEADOWS, AND FORAGE (pages 179–200)

(By I. P. ROBERTS)

PART III

THE ANIMAL, AND STOCK

CHAPTER XIII

THE OFFICES OF THE ANIMAL (pages 201–207)

CHAPTER XIV

How the Animal Lives (pages 208–238)

(By James Law, Director of the New York State Veterinary College, Cornell University)

CHAPTER XV

The Feeding of the Animal (pages 239–257)

(By H. H. Wing, Assistant Professor of Animal Industry in the Cornell University)

CHAPTER XVI

The Management of Stock (page 258-278)

(By I. P. Roberts)

THE PRINCIPLES OF AGRICULTURE

INTRODUCTION

1. *What Agriculture Is*

1. Agriculture, or farming, is the business of raising products from the land. These products are of two classes: crops, or plants and their products; stock, or animals and their products. The former are direct products of the land; the latter are indirect products of the land.

2. Agriculture also comprises, to a certain extent, the marketing or selling of its products. As marketable commodities, the products are of two classes: primary, or those which are put on the market in their native or natural condition, as wheat, potatoes, bananas, eggs, milk, wool; secondary, or those which are put on the market in a manufactured condition, as butter, cheese, cider, evaporated fruits.

3. The chief contribution of agriculture to the wealth and welfare of the world is the production of food. Its second contribution is the production of materials for clothing. Its third

is the production of wood or timber, used in building and in the various wood-working trades. Other contributions are the production of materials used in medicine and in various secondary and incidental arts and manufactures.

4. The ideal agriculture maintains itself. That is, it is able to thrive forever on the same land and from its own resources. The land becomes more productive with time, and this even without the aid of fertilizing materials from the outside. This state is possible only with a mixed husbandry, in which rotations of crops and the raising of animals are necessary features. The more specialized any agricultural industry becomes, the more must it depend upon outside and artificial aids for the enrichment of the land and for its continued support.

5. Agriculture may be roughly divided into four general branches or departments: agriculture in its restricted sense, animal industry, forestry, horticulture.

6. Agriculture in its restricted sense—sometimes, but erroneously, called agriculture proper—is a term applied to the general management of lands and farms, and to the growing of the staple grain and fiber crops. In North America, the use of the term agriculture has been restricted to the above application largely through the influence of agricultural colleges and experiment

stations, in which the general field of agriculture has been divided into various special subjects.

7. Animal industry is the raising of animals, either for direct sale or use or for their products. It is customary to speak of it as comprising three departments: stock-raising, or the general growing of mammals, as cattle, horses, sheep; dairy husbandry, or the production of milk and milk products; poultry-raising, or the growing of fowls, as chickens, turkeys, geese, ducks. In its largest sense, it comprises other departments, as apiculture or bee-raising, fish-culture, ostreaculture or oyster-raising, and the like.

8. Forestry is the growing of timber and woods. Its objects are two: to obtain a salable product; to produce some secondary effect upon the region, as the modification of climate or the preservation of the water-supply to rivers and lakes.

9. Horticulture is the growing of fruits, kitchen-garden vegetables, and ornamental plants. It has been divided into four departments: pomology, or fruit-growing; olericulture, or vegetable-gardening; floriculture, or the growing of flowers and plants for their own or individual uses as means of ornament; landscape horticulture, or the growing and planting of ornamental plants for their uses in mass effects in the landscape (on the lawn).

2. *The Personal Factors Upon Which Its Success Depends*

2*a*. *Upon business or executive ability*

10. Since the farmer makes a living by means of trade, it follows that ability to manage business and affairs is requisite to his success. Executive ability is as needful to him as to the merchant or the manufacturer; and the lack of such ability is probably the commonest and most serious fault with our agriculture. As the conditions of trade are ever changing, so the methods of the farmer must be amenable to modification. He must quickly and completely adapt himself to the commerce of the time. Manifestly, however, this business capability cannot be taught by books. It is a matter of temperament, home training, and opportunity. Like all permanent success, business prosperity depends upon correct thinking, and then upon the correct application of the thinking. Successful agriculture, therefore, is a matter of personality more than of circumstances.

11. The compound result of executive ability and experience may be expressed in the term farm-practice. It is the judgment of the farmer upon the question in hand. However much he may learn from science, his own

experience on his own farm must tell him what crops to grow, how to fertilize his land, what breeds and varieties to raise, when and how to sow and to reap. The experience of one farmer is invaluable to another, but each farm is nevertheless a separate and local problem, which the farmer must think out and work out for himself.

12. The farmer must be able not only to raise his products, but also to sell them. He must produce either what the trade demands, or be able to sell products which are not known in the general market. In other words, there are two types of commercial effort in farming: growing the staple products for the world's markets (as wheat, beans, maize, meat), in which case the market dictates the price; growing special products for particular or personal sale (as the products of superior excellence, and luxuries), in which case the producer looks for his customers and dictates the price.

2b. *Upon a knowledge of natural science*

13. The farmer, however, has more problems to deal with than those connected with trade. He must raise products: and such production depends upon the exercise of much special knowledge and skill. The most successful pro-

duction of agricultural products rests upon the application of many principles and facts of natural science; and the importance of such application is rapidly increasing, with the competitions and complexities of civilization. The study of these natural sciences also establishes habits of correct thinking, and opens the mind to a larger enjoyment of life,—for happiness, like success, depends upon habits of thought. The farmer should live for himself, as well as for his crops. The sciences upon the knowledge of which the best agricultural practice chiefly depends may now be mentioned, being stated approximately in the order of their importance to the actual practice of the modern farmer.

14. Physics. The physical properties and actions of bodies are fundamentally concerned in every agricultural result, whether the farmer knows it or not. The influences of light and heat, the movements of fluids in soil, plant and animal, the forces concerned in every machine and appliance, are some of the most obvious of these physical problems. So important to the farmer is a knowledge of physics that "agricultural physics" is now a subject of instruction in colleges. The most important direct application of a knowledge of physics to agricultural practice has come as a result of recent studies of the soil. The questions of soil moisture, soil

texture, the tilling of land, and the acceleration of chemical activities in the soil, are essentially questions of physics; and these are the kinds of scientific problems which the farmer needs first to apprehend.

15. Mechanics. In practice, mechanics is an application of the laws of physics. The elementary principles of mechanics are apprehended by the farmer unconsciously, as a result of experience; but since modern agriculture is impossible without numerous and often elaborate mechanical devices, it follows that it is not enough that the farmer be self-taught. At every turn the farmer uses or applies physical forces, in tools, vehicles, and machines. His work often takes him into the field of civil engineering. To show how much the farmer is dependent on practical mechanics, we need mention only implements of tillage, problems associated with the draughts of horse tools, the elaborate harvesting machinery, threshers and feed-mills and milk-working machinery and the power to run them, fruit evaporating machinery, pumps, windmills, hydraulic rams, construction of water supplies, problems of animal locomotion.

16. Plant-knowledge, or botany. Since the plant is the primary product of the farm, a knowledge of its characteristics and kinds is of

fundamental importance to the farmer. From the farmer's standpoint, there are four great departments of plant-knowledge: physiology, or a knowledge of the way in which the plant lives, grows, and multiplies; pathology, or a knowledge of mal-nutrition and diseases; systematic botany, or a knowledge of the kinds of plants; ecology, or a knowledge of the inter-relations between plants and their environments (or surroundings), and how they are modified by changes in environments, by crossing, and by breeding.

17. Animal-knowledge, or zoölogy. There are also four general directions in which animal-knowledge appeals to the farmer: physiology, with its practical applications of feeding, housing, and general care of animals; pathology, or knowledge of mal-nutrition and diseases (with special applications in the practice of surgery and medicine); kinds of animals, and the life-histories of those which are particularly beneficial or injurious to agriculture (with special applications in economic entomology and economic ornithology); ecology and breeding.

18. Chemistry. There are two general directions in which chemistry appeals to the agriculturist: in enlarging his knowledge of the life-processes of plants and animals; and in affording direct information of the composition

of many materials used or produced on the farm. In practice, chemistry aids the farmer chiefly in suggesting how he may feed plants (fertilize the land) and animals. So many and important are the aids which chemistry extends to agriculture, that the various subjects involved have been associated under the name of "agricultural chemistry." This differs from other chemistry not in kind, but only in the subjects which it considers.

19. Climatology. Climate determines to a large extent the particular treatment or care which the farmer gives his crops and stock. It also profoundly influences plants and animals. They change when climate changes, or when they are taken to other climates. Climate is therefore a powerful agency in producing new breeds and new varieties. The science of weather, or meteorology, is also intimately associated with the work of the farmer.

20. Geology. The agricultural possibilities of any region are intimately associated with its surface geology, or the way in which the soil was formed. A knowledge of the geology of his region may not greatly aid the farmer in the prosecution of his business, but it should add much interest and zest to his life.

21. We now apprehend that agriculture is a complicated and difficult business. Founded

upon trade, and profoundly influenced by every commercial and economic condition, its successful prosecution nevertheless depends upon an intimate and even expert knowledge of many natural sciences. Aside from all this, the farmer has to deal with great numbers of objects or facts: thousands of species of plants are cultivated, and many of these species have hundreds and thousands of varieties; many species of animals are domesticated, and each species has distinct breeds. Each of these separate facts demands specific treatment. Moreover, the conditions under which the farmer works are ever changing: his innumerable problems are endlessly varied and complicated by climate, seasons, vagaries of weather, attacks of pests and diseases, fluctuations in labor supply, and many other unpredictable factors.

3. *Its Field of Production*

22. In the production of its wealth, agriculture operates in three great fields,—with the soil, the plant, and the animal. Although aided at every point by knowledge of other subjects, its final success rests upon these bases; and these are the fields, therefore, to which a textbook may give most profitable attention.

SUGGESTIONS ON THE FOREGOING PARAGRAPHS .

1*a*. The word agriculture is a compound of the Latin *agri*, "field," and *cultura*, "tilling." Farming and husbandry are synonymous with it, when used in their broadest sense; but there is a tendency to restrict these two words to the immediate practice, or practical side, of agriculture.

2*a*. It is often difficult to draw a line of demarkation between agriculture and manufacture. The husbandmen is often both farmer and manufacturer. Manufacturing which is done on the farm, and is of secondary importance to the raising of crops or stock, is commonly spoken of as agriculture. The manipulation or manufacturing of some agricultural products requires such special skill and appliances that it becomes a business by itself, and is then manufacture proper. Thus, the making of flour is no longer thought of as agriculture; and the making of wine, jellies, cheese, butter, canned fruits, and the like, is coming more and more into the category of special manufacturing industries. Strictly speaking, agriculture stops at the factory door.

3*a*. Agriculture is often said to be the most fundamental and useful of occupations, since it feeds the world. Theoretically, this may be true; but a high state of civilization is possible only with diversification of interests. As civilization advances, therefore, other occupations rise in relative importance, the one depending upon the other. In our modern life, agriculture is impossible without the highly developed manufacturing and transportational trades. Broadly speaking, civilization may be said to rest upon agriculture, transportation, and manufacture.

4*a*. Mixed husbandry is a term used to denote the growing of a general variety of farm crops and stock, especially the growing of grass, grain, with grazing (pasturing) and general stock-raising. It is used in distinction to specialty-farming or the raising of particular or special things, as fruit, bees, vegetables, beef, eggs.

4*b*. Self-perpetuating industries conduce to stability of political and social institutions. "The epochs which precede the agricultural occupation of a country are commonly about as follows: Discovery, exploration, hunting, speculation, lumbering or mining. The real and permanent prosperity of a country begins when the agriculture has evolved so far as to be self-sustaining and to leave the soil in constantly better condition for the growing of plants. Lumbering and mining are simply means of utilizing a reserve which nature has laid by, and these industries are, therefore, self-limited, and are constantly moving on into unrobbed territory. Agriculture, when at its best, remains forever in the same place, and gains in riches with the years; but in this country it has so far been mostly a species of mining for plant-food, and then a rushing on for virgin lands."—*Principles of Fruit-Growing, 26.*

8*a*. Forestry is popularly misunderstood in this country. The forest is to be considered as a crop. The salable product begins to be obtainable in a few years, in the shape of trimmings and thinnings, which are useful in manufacture and for fuel; whereas, the common notion is that the forest gives no return until the trees are old enough to cut for timber. One reason for this erroneous impression is the fact that wood has been so abundant and cheap in North America that the smaller products have not been considered to be worth the saving; but even now, in the manufacture of various articles of commerce, the trimmings and thinnings of forests should pay an income on the investment in some parts of the country. If a manipulated forest is a crop, then forestry is a kind of agriculture, and it should not be confounded with the mere botany of forest trees, as is commonly done.

9*a*. The word horticulture is made up of the Latin *hortus*, "garden," and *cultura*, "tilling." In its broadest sense, the word garden is its equivalent, but it is commonly used to designate horticulture as applied to small areas, more particularly when the subjects are flowers and kitchen-garden vegetables. Etymologically, *garden* refers to the engirded or confined (walled-in or fenced-in) area immediately surrounding the residence, in distinction to the *ager* (1*a*) or field which lay beyond. *Hortus* has a similar significance. *Paradise* is, in etymology, a name for an enclosed area; and the term was given to some of the early books on gardening, e. g., Parkinson's "Paradisus Terrestris" (1629), which is an account of the ornamental plants of that period.

14*a*. King's book on "The Soil" explains the intimate relation of physical forces to the productivity of the land; and the author is Professor of Agricultural Physics in the University of Wisconsin. There is a Division of Soils in the National Department of Agriculture, the work of which is largely in the field of soil physics. The physical or mechanical analysis of soils is now considered to be as important as the chemical analysis. Some of the physical aspects of farm soils are discussed in our chapters ii., iii., iv., v.

16*a*. Ecology (written œcology in the dictionaries) is the science which treats of the relationship of organisms (that is, plants and animals) to each other and to their environments. It is animal and vegetable economy, or the general external phenomena of the living world. It has to do with modes and habits of life, as of struggle for existence, migrations and nesting of birds, distribution of animals and plants, influence of climate on organisms, the way in which any plant or animal behaves, and the like. Darwin's works are rich in ecological observations.

16*b*. Environment is the sum of conditions or surroundings or circumstances in which any organism lives. An environment of any plant is the compound condition produced by soil, climate, altitude, struggle for existence, and so on.

18*a*. It is customary to consider agricultural chemistry as the fundamental science of agriculture. Works on agricultural chemistry are often called works on agriculture. But agriculture has no single fundamental science. Its success, as we have seen, depends upon a union of business methods and the applications of science; and this science, in its turn, is a coördination of many sciences. Chemistry is only one of the sciences which contribute to a better agriculture. Under the inspiration of Davy, Liebig, and their followers, agricultural chemistry made the first great application of science to agriculture; and upon this foundation has grown the experiment-station idea. It is not strange, therefore, that this science should be more intimately associated than others with agricultural ideas; but we now understand that agriculture cannot be an exact or definite science, and that the retort and the crucible can solve only a few of its many problems. In particular, we must outgrow the idea that by analyzing soil and plant we can determine what the one will produce and what the other needs. Agricultural chemistry is the product of laboratory methods. The results of these methods may not apply in the field, because the conditions there are so different and so variable. The soil is the laboratory in which the chemical activities take place, but conditions of weather are ever modifying these activities; and it is not always that the soil and the plant are in condition to work together.

20*a*. As an illustration of the agricultural interest which attaches to the surface geology of a region, see Tarr's "Geological History of the Chautauqua Grape Belt," Bull. 109 Cornell Exp. Sta.

21*a*. Probably no less than 50,000 species of plants (or forms which have been considered to be species) have been cultivated. The greater number of these are ornamental subjects. Of orchids alone, as many as 1,500 species have been introduced into cultivation. Nicholson's Illustrated Dictionary of Gardening describes about 40,000 species of domesticated plants. Of plants grown for food, fiber, etc., De Candolle admits 247 species (in Origin of Cultivated Plants), but these are only the most prominent ones. Vilmorin (The Vegetable Garden) describes 211 species of kitchen-garden vegetables alone. Sturtevant estimates (Agricultural Science, iii., 178) 1,076 species as having been "recorded as cultivated for food use." Of some species, the cultivated varieties are numbered by the thousands, as in apple, chrysanthemum, carnation, potato. Of animals, more than 50 species are domesticated, and the breeds or varieties of many of them (as in cattle) run into the hundreds.

21*b*. It is commonly said that agriculture is itself a science, but we now see that this is not true. It has no field of science exclusively its own. Its purpose is the making of a living for its practitioner, not the extension of knowledge. The subject of mathematics is numbers, quantity and magnitude; of botany, plants; of ornithology, birds; of entomology, insects; of chemistry, the composition of matter; of astronomy, the heavens: but agriculture is a mosaic of many sciences, arts and activities. Or, it may be said to be a composite of sciences and arts, much as medicine and surgery are. But if there is no science of agriculture as distinct from other sciences, the prosecution of agriculture must be scientific; and the fact that it is a mosaic makes it all the more difficult to follow, and enforces the importance of executive judgment and farm-practice over mere scientific knowledge.

22*a*. The province of a text-book of agriculture, in other words, is to deal (1) with the original production of agricultural wealth rather than with its manufacture, transportation or sale, for these latter enterprises are largely matters of personal circumstance and individuality, and (2) with those principles and facts which are common to all agriculture, or which may be considered to be fundamental.

22*b*. In other words, we must search for principles, not for mere facts or information: we shall seek to ask why before we ask how. Principles apply everywhere, but facts and rules may apply only where they originate. Agriculture is founded upon laws; but there are teachers who would have us believe that it is chiefly the overcoming of mere obstacles, as insects, unpropitious weather, and the like. There are great fundamentals which the learner must comprehend; therefore we shall say nothing, in this book, about the incidentals, as the kinds of weeds, the brands of fertilizers, the breeds of animals, the varieties of flowers.

PART I

THE SOIL

CHAPTER I

THE CONTENTS OF THE SOIL

1. *What the Soil Is*

23. The earth, the atmosphere, and the sunlight are the sources of all life and wealth. Atmosphere and sunlight are practically beyond the control of man, but the surface of the land is amenable to treatment and amelioration.

24. The soil is that part of the solid surface of the earth in which plants grow. It varies in depth from less than an inch to several feet. The uppermost part of it is usually darkest colored and most fertile, and is the part which is generally understood as "the soil" in common speech, whereas the under part is called the sub-soil. When speaking of areas, we use the word land; but when speaking of the particular agricultural attributes of this land, we may use the word soil.

2. *How Soil Is Made*

2*a*. *The inorganic elements*

25. The basis of soil is fragments of rock. To this base is added the remains of plants and animals (or organic matter). When in condition to grow plants, it also contains water. The character of any soil, therefore, is primarily determined by the kind of rock from which it has come, and the amount of organic matter and water which it contains.

26. As the surface of the earth cooled, it became rock-bound. Wrinkles and ridges appeared, forming mountains and valleys. The tendency is for the elevations to be lessened and the depressions to be filled. That is, the surface of the earth is being leveled. The chief agency in this leveling process is weathering. The hills and mountains are worn down by alternations of temperature, by frost, ice, snow, rain and wind. They are worn away by the loss of small particles: these particles, when gathered on the hillsides or deposited on lower levels, form soil.

27. The weathering agencies which reduce the mountains operate also on level areas; but since the soil then remains where it is formed, and thereby affords a protection to the underlying rock, the reduction of the rock

usually proceeds more slowly than on inclined surfaces.

28. There are, then, two sets of forces concerned in the original formation of soils,—the disintegration or wearing away of the rock, and the transfer or moving of the particles to other places.

2b. *The organic elements and agents*

29. Plants are important agents in the formation of soil. Their action is of two kinds: the roots corrode and break up the surfaces of rock and particles of soil, and the plant finally decays and adds some of its tissue to the soil.

30. In the disintegration of rock and the fining of soil, the root acts in two ways: it exerts a mechanical force or pressure as it grows, cracking and cleaving the rock; and it has a chemical action in dissolving out certain materials, and thereby consuming and weakening the rock.

31. Animals contribute to the formation of soil by their excrement and the decay of their carcasses. Burrowing and digging animals also expose rocks and soils to weathering, and contribute to the transportation of the particles. Some animals are even more directly concerned in soil-making. Of these, the chief are the

various kinds of earthworms, one of which is the common angleworm. These animals eat earth, which, when excreted, is more or less mixed with organic matter, and the mineral particles are ground and modified. It is now considered that in the tenacious soils in which these animals work, the earthworms have been very important agents in fitting the earth for the growing of plants, and consequently for agriculture.

32. While the basis of most soils is disintegrated rock, there are some soils which are essentially organic in origin. These are formed by the accumulation of vegetable matter, often aided by the incorporation of animal remains. In the tropics, such soils are often formed on shores and in lagoons by the extension of the trunk-like roots of mangroves and other trees. In the network of roots, leaves and sea-wrack are caught, and mold is formed. Water plants (as marsh grasses and eel-grass) are sometimes so abundant on sea margins as to eventually form solid land. On the edges of lakes and ponds, the accumulation of water-lily rhizomes and other growths often affords a foothold for sedges and other semi-aquatic plants; and the combined growth invades the lake and often fills it. Portions of this decaying and tangled mass are sometimes torn away by wind

or wave, and become floating islands. Such islands are often several acres in extent. In high latitudes, where the summer's growth does not decay quickly, one season's growth is sometimes added above another until a deep organic soil is formed. This is especially noticeable in the gradual increase in height of sphagnum swamps. Peat bogs are organic lands, and they fill the beds of former lakes or swamps. Of course, all these organic soils contain mineral matter, but it is mostly such as comes from the decay of the plants themselves. It was originally obtained from the earth, but is used over and over again; and each year a little new material may be added by such plants as reach into the hard land below, and by that which blows into the area in dust.

33. Decaying organic matter forms mold or humus. The mineral elements may be said to give "body" to the soil, but the humus is what gives it "life" or "heart." Humus makes soils dark-colored and mellow. Humus not only adds plant-food to the soil, but improves the physical condition of the soil and makes it congenial for plants. It augments the water-holding capacity of the soil, modifies the extremes of temperature, facilitates the entrance of air, and accelerates many chemical activities. It is the chief agent in the formation of loam:—a sandy loam is a

friable soil rich in vegetable matter, the original basis of which is sand; a clay loam is one similarly ameliorated, the basis of which is clay. "Worn-out" lands usually suffer more from lack of humus than from lack of actual plant-food, and this explains why the application of stable-manure is so efficacious.

34. There are three general ways in which humus is obtained in farm-practice: (1) By means of the vegetable matter which is left on or in the ground after the crop is removed (as roots, stubble, sod, garden refuse); (2) by means of crops grown and plowed under for that particular purpose (green-manuring); (3) by means of direct applications to the land (as compost and stable-manure). The deeper and more extensive the root-system of any plant, the greater, in general, is its value as an ameliorator of soil, both because it itself exerts a more wide-spread influence (30), and because when it decays it extends the ameliorating effects of humus to greater depths.

35. Aside from these varied component elements, fertile soil is inhabited by countless numbers of microscopic organisms, which are peculiar to it, and without which its various chemical activities can not proceed. These germs contribute to the breaking down of the soil particles and to the decay of the organic materials, and

in doing so, aid in the formation of plant-foods. The soil, therefore, is not merely an inert mass, operated upon only by physical and chemical forces, but it is a realm of intense life; and the discovery of this fact has radically modified our conception of the soil and the means of treating it. Enriching the land is no longer the adding of mere plant-food: it is also making the soil congenial to the multiplication and well-being of micro-organisms.

2c. *Transportation of soils*

36. The soil is never at rest. The particles move upon each other, through the action of water, heat and cold, and other agencies. The particles, whether of inorganic or organic origin, are also ever changing in shape and composition. They wear away and crumble under the action of weather, water, organic acids of the humus, and the roots of plants. No particle of soil is now in its original place. These changes are most rapid in tilled lands, because the soil is more exposed to weather through the tillage and the aerating effect of deep-rooted plants (as clover); and the stirring or tilling itself wears the soil particles. Even stones and pebbles wear away (26*a*); and the materials which they lose usually become productive elements of the

soil. Some lands have very porous or "rotten" stones, and these pass quickly into soil. Stones are no doubt a useful reserve force in farm lands, giving up their fertility very gradually, and thereby saving some of the wastefulness of careless husbandry. The general tendency, in nature, is for soils to become finer, more homogeneous, and better for the growth of plants.

37. But there are greater movements than these. Soil is often transported long distances, chiefly by means of three agents: moving water, ice and snow, wind. Transported soils are apt to be very unlike the underlying rock (or original surface), and they are often very heterogeneous or conglomerate in character. Soils which remain where they are formed (27) naturally partake of the nature of the bedrock, and are generally more homogeneous than transported soils, as, for example, the limestone soils which overlie great deposits of lime-rock.

38. Moving water always moves land. The beating of waves wears away rocks and stones and breaks up debris, and deposits the mass on or near the shore. Streams carry soils long distances. The particles may be in a state of suspension in the water, and be precipitated in the quieter parts of the stream or in bayous or lagoons, or they may be driven along the bed of the stream by the force of the current, and

be deposited wherever obstructions occur, or be discharged on the delta at the mouth. The deposition of sediment in times of overflow adds new vigor to the submerged lands. The historic example of this is the Nile valley, but all bottom lands which are subjected to periodical overflows teach the same lesson. Alluvial lands are formed from the deposition of the sediment of water.

39. In mountainous regions, snow and ice carry away great quantities of rock and soil. The most powerful transporters of soil are glaciers, or moving masses of ice. Glaciers loosen the rock and then grind and transport it. In the glacial epoch, in which much of the northern part of the northern hemisphere was covered with gigantic ice-sheets slowly moving to the southward, enormous quantities of rock and earth were transported, and deposited wherever the ice melted. In eastern North America, the ice-front advanced to the latitude of the Ohio river, and the boulder-strewn fields and varied soils to the northward of this latitude are the legacy which the epoch left to the farmer.

40. In all areas which are subjected to periods of drought, the wind transports soils in the form of dust, often in great amounts and for long distances. In some parts of the world, so much earth is carried by violent winds that

these winds are known as "sand-storms." Most shores, particularly if sandy, are much modified by the action of wind. But the wind has an influence upon soils even in the most protected and equable regions. The atmosphere contains dust, much of which is valuable plant-food. This dust is transported by winds, and it finally settles or is carried down by snow and rain. Although the amount of dust which is deposited in any given time may be slight, it is nevertheless continuous, and has an important effect upon the soil.

3. *The Resources of the Soil*

41. The soil affords a root-hold for plants,—a place in which they can grow. It also supplies the environmental conditions which roots need,—protection, moisture, air, agreeable temperature, and other congenial surroundings.

42. The soil is also a store-house of plant-food. Roberts calculates, from many analyses, that in average agricultural lands the surface eight inches of soil on each acre contains over 3,000 pounds of nitrogen, nearly 4,000 pounds of phosphoric acid, and over 17,000 pounds of potash. These three elements are the ones which the farmer must chiefly consider in maintaining or augmenting the productive power of the land ;

yet the figures "reveal the fact that even the poorer soils have an abundance of plant-food for several crops, while the richer soils in some cases have sufficient for two hundred to three hundred crops of wheat or maize." Yet these calculations are made from only the upper eight inches of soil.

43. Happily, this food is not all directly available or useful to plants (being locked up in insoluble combinations), else it would have been exhausted by the first generations of farmers. It is gradually unlocked by weather, micro-organisms, and the roots of plants; and the better the tillage, the more rapid is its utilization. Plants differ in the power to unlock or make use of the fertility of the soil.

44. Nature maintains this store of fertility by returning her crops to the soil. Every tree of the forest finally crumbles into earth. She uses the materials, then gives them back in a refined and improved condition for other plants to use. She repays, and with interest.

45. Man removes the crops. He sends them to market in one form or another, and the materials are finally lost in sewage and the sea. He sells the productive power of his land; yet it does not follow that he impoverishes his soil in proportion to the plant-food which he sells. Given the composition of any soil and

of the crops which it is to produce, it is easy to calculate the time when the soil will have lost its power; but it must be remembered that the materials which the plant removes are consumed, and that the volume of the soil is reduced by that amount. The result is, therefore, that the deeper parts of the soil are brought into requisition as fast as the upper parts are consumed; and these depths will last as long as the earth lasts.

46. Of some materials, however, the plant uses more freely than of others, in proportion to their abundance in the soil. Therefore the soil may finally lose its productivity, although it is doubtful if it can ever be completely exhausted of plant-food.

47. Again, the profit in agriculture often lies in making the soil produce more abundantly than it is of itself able to do. That is, even after tillage and every other care have forced the soil to respond to its full ability, it may pay the farmer to buy plant-food in bags in the same way that it may pay him to buy ground feed when fattening sheep. Whether it is advisable to buy this plant-food is a matter of business judgment which every farmer must determine for himself, after having considered the three fundamental factors in the problem: the cost of the plant-food (or fertilizer), the

probable effect of this extra food upon the crop, and the commercial value of the extra crop. In general, it should be considered that in mixed husbandry the fertility of the land must be maintained by means of farm-practice (that is, by good farming), and that plant-food should be bought only for the purpose of producing the extra product.

48. We are now able to comprehend that the soil is a compound of numberless inorganic and organic materials, a realm of complex physical and chemical forces, and the scene of an intricate round of life. We must no longer think of it as mere dirt. Moreover, we are only beginning to understand it; and if the very soil is unknown to us, how complicated must be the great structure of agriculture which is reared upon it!

SUGGESTIONS ON CHAPTER I

25a. The word organic refers to animals and plants or their products and remains; that is, to things which live and have organs. Organic compounds, in chemistry, are those which have been built up or produced by the action of a plant or animal. Modern usage, however, defines organic compounds as those which contain carbon. Starch, sugar, woody fiber, are examples.

25b. Inorganic compounds are such as are not produced by living organisms, as all the mineral compounds. They are found in the earth and air. Salt, potash, phosphoric acid, lime, are examples.

25*c*. The organic matter in soils—the plant and animal remains—is removed by burning. Let the pupil secure a cupful of wet soil and carefully weigh it on delicate scales. Then let it dry in the sun, and weigh again ; the difference in weight is

Fig. 1. Showing the wearing away of mountain peaks and the formation of soil at the base.

due to the loss or evaporation of water. Now place it in a moderately warm oven or on a stove, and after a few minutes weigh again ; more of the water will now have passed off. Now thoroughly burn or bake it, and weigh ; the loss is now mostly due to the burning of the organic matter, and part of this matter has passed off as gas. If there is no perceptible loss from the burning, it is evidence that the sample contained little organic matter. Note the difference in results between clay and muck. The pupil may also be interested to try to grow plants in the baked soil.

26*a*. The wearing away of rock by the weather may be observed wherever stones are exposed. Even granite and marble monuments lose their polish and luster in a few years. The sharp and angular projections disappear from the ledges and broken stones of railway cuts and quarries. The pupil should look for the wear on any rocks with which he may be familiar. All stones tend to grow smaller. On a large scale, the wasting of rocks may be seen in the debris at the base of precipices and mountain peaks (Fig. 1), or wherever steep walls of rock are exposed. The palisades of the Hudson, and other precipitous river and lake bluffs, show this action well. Mountains tend to become rounded in the long processes of time, although some rocks are of such structure that they hold their pointed shape until worn almost completely away. In Geikie's "Geological Sketches," Essay No. 8, the reader will find an interesting account of weathering as illustrated by the decay of tombstones.

26*b*. The extent of this weathering and denuding process in the formation of soils may be graphically illustrated by the present conformation of the Alps and adjacent parts of Europe. Lubbock writes that "much of the deposits which occupy the valleys of the Rhine, Po, Rhone, Reuss, Inn, and Danube—the alluvium which forms the plains of Lombardy, of Germany, of Belgium, Holland, and of southeast France—consists of materials washed down from the Swiss mountains." The amount of material which has been removed from the Alps is probably "almost as great as that which still remains." So great has been the denudation that in certain cases "what is now the top of the mountain was once the bottom of a valley." The Matterhorn, the boldest and one of the highest of the Alps, "is obviously a remnant of an ancient ridge," and the "present configuration of the surface [of Switzerland] is indeed mainly the result of denudation. * * It is certain that not a fragment of the original surface is still in existence, though it must not be inferred that the mountains were at any time so much higher, as elevation and denudation went on together." There is even evidence to show that an earlier range of mountains occupied the site of the present Alps, and that these old mountains were removed or worn away by denudation.—*See Sir John Lubbock, "Scenery of Switzerland," Chaps. iii. and iv.*

29*a*. Even hard surfaces of rock often support lichens, mosses, and other humble plants. "The plant is co-partner with the weather in the building of the primal soils. The lichen spreads its thin substance over the rock, sending its fibers into the crevices and filling the chinks, as they enlarge, with the decay of its own structure; and finally the rock is fit for the moss or fern or creeping vine, each newcomer leaving its impress by which some later newcomer may profit. Finally the rock is disintegrated and comminuted, and is ready to be still further elaborated by corn and ragweed. Nature intends to leave no vacant or bare places. She providently covers the railway embankment with quack-grass or willows, and she scatters daisies in the old meadows where the land has grown sick and tired of grass." —*Principles of Fruit-Growing, 176.*

Fig. 2. The halves of a rock forced apart by the growth of a tree.

30*a*. It is interesting to consider the general reasons for the evolution of the root. Plants were at first aquatic, and probably absorbed food from the water on all their surfaces. They may not have been attached to the earth. As they were driven into a more or less terrestrial life by the receding of the waters and as a result of the struggle for existence, they developed parts which penetrated the earth. These parts were probably only hold-fasts at first, as the roots of many seaweeds are at the present time. But as it became less and less possible for the general surface of the plant to absorb food, the hold-fast gradually became a food-gathering or feeding member.—*See Survival of the Unlike, pp. 41–43.*

30*b*. If the pupil has access to ledges of rock on which trees are growing, he will readily be able to satisfy himself that roots force open cracks and thereby split and sever the stone. Fig. 2 is an example, showing how a black cherry tree, gaining a foothold in a crevice, has gradually forced the parts of the rock

Fig. 3. Lichens have obtained a foothold.

asunder. This particular example is the "half-way stone" between the Michigan Agricultural College and the city of Lansing. Fig. 3 shows a stone upon which lichens have obtained a foothold. Any person who has worked much in a garden will have seen how roots often surround a bone, taking their food from its surface and insinuating themselves into the cracks. Roots will corrode or eat out the surface of marble. On a polished marble block place a half inch of sawdust, in which plant seeds. After the plants have attained a few leaves, turn the mass of sawdust over and observe the prints of the roots on the marble (Fig. 4).

30c. By chemical action is meant the change from which results a new chemical combination. It produces a rearrangement of molecules. For example, the change which takes place when, by uniting lime and sulfuric acid, sulfate of lime or gypsum is produced, is chemical action.

31*a*. Knowledge of the work of the earthworm in building soils dates practically from the issue of Darwin's remarkable book, "The Formation of Vegetable Mould, through the Action of Worms," which the reader should consult for particulars. The subject is also considered briefly in King's "Soil," Chap. i., which also discusses the general means of soil-building.

32*a*. As an example of the formation of organic soils in the tropics, read accounts of the mangrove. Its mode of propagation

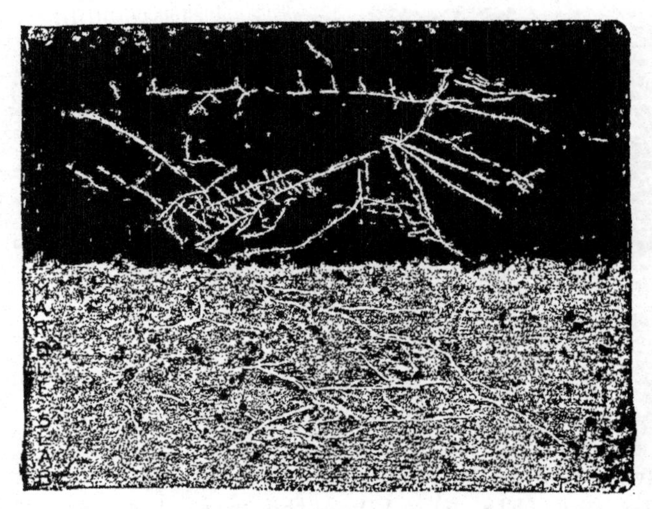

Fig. 4. The etching of marble by roots. The upper part represents the bottom of the sawdust soil, turned back. The prints of the roots are seen in the marble below.

is explained, with illustrations, in Bailey's "Lessons with Plants," pp. 371–374; the tree is also described in Chap. v. of Gaye's "Great World's Farm." As an example of a formation of a peat bog by the growth of sphagnum, read Ganong "On Raised Peat-bogs in New Brunswick," Botanical Gazette, pp. 123–126, May, 1891. Sphagnum is moss which grows in cold bogs. Nurserymen and florists use it in the packing of plants.

33*a*. When spelled *humus*, the word is a noun; when spelled *humous*, it is an adjective, as "humous soils."

34*a*. Compost is decayed or decaying organic matter which it is intended shall be applied to the land. It is usually obtained by placing leaves, sod, manure or litter in a low flat-topped (so that it will catch the rain) pile, and "turning it," or forking it

Fig. 5. A compost pile.

over, every few weeks, to prevent heating and to hasten uniform decomposition (Fig. 5). When the mass has passed into the condition of humus or mold (or become fine and soil-like), it is applied to the land. Composting is a most useful means of utilizing leaves, garden refuse, and other materials which are too coarse or "raw" to be applied directly to the land.

35*a*. "The term micro-organism is a general one, which includes any very minute, microscopic form of life. More strictly speaking, the word has come to apply especially to certain forms of plant life which are too minute to be seen individually by the naked eye, and which hence require for their study the higher powers of the microscope."—*Fred'k D. Chester, Bull. xl., Del.*

Exp. Sta. The terms *germ*, *microbe*, *bacterium* (plural *bacteria*), are popularly used in the same sense as micro-organism. These beings are usually unicellular (each one consisting of only a single cell). They are generally classified with plants. The rôle of micro-organisms in rendering soil elements available to plants is very complex and not yet well understood. Popular discussions of the subject, by Chester, will be found in 10 Ann. Rep. and Bull. xl. of Del. Exp. Sta. The relation of germs to nitrification is briefly discussed in King's "Soil," pp. 125–134, and Roberts' "Fertility," 244–248. Fig. 6 illustrates one of the common bacteria, very much magnified. This species (*Bacillus ubiquitus*) is abundant in water, air, and decaying substances.

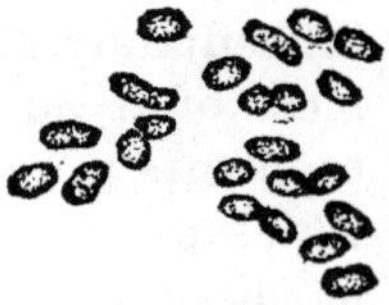

Fig. 6. Micro-organisms, greatly magnified.

38*a*. Observe the deposits of sand in the quiet side (usually the concave side) of streams, and also the delta where a rapid rill flows into a slow one. When the rill flows into a rapid stream, the larger current carries away the deposit so that it may not be seen. Recall how sand-bars form again and again in lakes, and how streams must be frequently dredged to keep the channel open. The slower the stream the more quickly does it drop its sediment; and the more winding, also, is its course, lying in the bed of its own deposits.

38*b*. Dip a glass of water from a roily stream, and observe the earth which settles to the bottom.

39*a*. Glaciers are still abundant in alpine and arctic regions. It was from the study of glaciers in the Alps that Agassiz conceived the hypothesis that large parts of the earth had once been subjected to glacial action. A good popular discussion of glaciers and their action may be found in Chap. xvii. of Tarr's "Elementary Physical Geography." Delightful readings may also be made from Agassiz's "Geological Sketches."

40*a*. Let the pupil catch a few rain drops on a perfectly clean and clear pane of glass, and observe if any sediment is left when the drops have evaporated. Is there any difference in the amount of dust brought down after a "dry spell" and after a period of rainy weather, or at the beginning and end of a shower? The pupil may now be able to explain why the windows get dirty after a rain; and he will be interested in the streaks on the cornices of buildings and on exposed statuary. He may have heard that even sailing ships get dusty when at sea.

42*a*. See Roberts' "Fertility of the Land," p. 16. Read all of Chapter i. The food which is not available, or not in condition to be used by the plant, but which may become available through good tillage or otherwise, is called *potential* plant-food.

43*a*. The soil is not a simple reservoir of plant-food in the condition of salt or sugar, ready to be dissolved in water and immediately taken up by roots. The soil is plant-food; but most of it must be changed in composition before it is available to plants; and the elements are not present in the proportions which plants require, so that much of the soil is in excess of the needs of plants and can never be used as food.

43*b*. "We asked the question why mullein can thrive on a piece of poor soil, and wheat cannot. Many replies were that the mullein took different plant-food than the wheat. This is not true. All plants require the same elements, although not always in the same proportions. The point we wish to make plain is that some plants can readily find these elements in soil, and that other plants cannot. Some plants, like the hardy goat, can find sustenance on rocky hillsides, and other plants, like the high-bred sheep, must have the food easily digestible, or, as is generally said, in available form."—*John W. Spencer.*

48*a*. For supplementary reading on the formation of soils, Chapter i. of King's "Soil" should be consulted. Most text-books of geology also treat the subject to some extent. Shaler's article on soil, in 12th Annual Report of the U. S. Geological Survey (pp. 319–345), is excellent. A discussion of weathering may be found in Chapter vi. of Tarr's "Elementary Geology;" and other references are contained in Chapters xiii. and xxi. of his "Elementary Physical Geography." Stockbridge's "Rocks and Soils" (1895) has special reference to agriculture. A readable account of the formation of soil may be found in Chapters iii., iv. and v., Gaye's "Great World's Farm." Merrill's "Rocks, Rock-Weathering and Soils" (1897) is a full scientific discussion of the subject.

Chapter II

THE TEXTURE OF THE SOIL

JOHN W. SPENCER

1. *What Is Meant by Texture*

49. We have seen that the offices of the soil are of two general kinds,—it affords a physical medium in which the plant can grow (41), and it supplies materials which the plant uses in the building of its tissues (42). It cannot be said that one of these offices is more important than the other, since both are essential; but attention has been so long fixed upon the mere contents of soils that it is important to emphasize the physical attributes. Crops cannot grow on a rock, no matter how much plant-food it may contain. The passing of rock into soil is a matter of change in texture more than in plant-food.

50. The physical state of the soil may be spoken of as its texture, much as we speak of the texture of cloth. The common adjectives which are applied to the condition of agricultural soils are descriptions of its texture: as,

mellow, hard, loose, compact, open, porous, shallow, deep, leachy, retentive, lumpy, cloddy, fine, in good tilth.

51. Texture must not be confounded with the physical forces or operations in the soil, as the fluctuations of temperature, movements of water, circulation of air. Texture refers to condition or state, and is passive, not to forces or movements, which are active; but it is upon this passive condition that the operation of both physical and chemical forces chiefly depends.

2. *Why Good Texture Is Important*

52. A finely divided, mellow, friable soil is more productive than a hard and lumpy one of the same chemical composition, because: It holds and retains more moisture; holds more air; promotes nitrification; hastens the decomposition of the mineral elements; has less variable extremes of temperature; allows a better root-hold to the plant; presents greater surface to the roots. In all these ways, and others, the mellowness of the soil renders the plant-food more available, and affords a congenial and comfortable place in which the plant may grow.

53. Good texture (as understood by the farmer) not only facilitates and hastens the physical and chemical activities, but it also presents

a greater feeding-surface to roots, because the particles of earth are very small (52). Roots feed on the surfaces of hard particles of earth, and the feeding-area is therefore increased in proportion to the increase in the surface area of the particles. Dividing a cube into two equal parts increases its surface area by one-third. (Dividing a cube adds two sides or surfaces.) Fining the soil may therefore be equivalent to fertilizing it, so far as plant-growth is concerned.

3. *How Good Texture Is Obtained*

54. In agricultural lands there are two general ways of improving the texture of soils,—by modifying the size of the particles, and by adding extraneous or supplementary materials.

55. Compact and lumpy soils are usually improved when the particles are made smaller: such soils then become "mellow." Very loose or leachy soils are usually improved if the particles, particularly in the under-soil or subsoil, are brought together and compacted: such soils then become "retentive" (that is, they hold more water and plant-food).

56. The size of the particles of soils is modified by three general means: (*a*) by applying mechanical force, as in all the operations of

tilling; (*b*) by setting at work various physical forces, as weathering (fall-plowing is a typical example), and the results following under-draining; (*c*) by applying some material which acts chemically upon the particles. (The first caption, *a*, is illustrated in paragraphs 26, 26*a*, 26*b*, 27, 28; and it is further explained in Chapter iv.)

57. (*b*) Under-drainage has two general uses, —it removes superfluous water, and improves the physical condition of the soil. The latter use is often the more important. The improvement of the texture is the result, chiefly, of preventing water-soaking and of admitting air. Under-drained soils become "deeper,"— that is, the water-table (or stratum, which is more or less impervious to water) is broken down or loosened. The water-table is said to be lowered, since the depth at which water stands tends to approach nearer and nearer to the depth of the drains.

58. (*c*) Some substances have the power to break down or to pulverize hard soils, or to bind together loose ones, or to otherwise modify the texture. Such materials—which are applied for their remote or secondary chemical effects—are called amendments. Lime is a typical example. Quick-lime is known to make clay lands mellow, and it is supposed to cement or bind

together the particles of sands or gravels. Most chemical fertilizers are both amendments and direct fertilizers, since they modify the texture of the soil as well as add plant-food to it.

59. The extraneous or supplementary materials (54) which directly modify the texture of soils are those which make humus (33), as green-manures, farm-manures, and the like. Stable-manure is usually more important in improving soil texture than in directly supplying plant-food.

4. *Texture and Manures*

60. We have now seen that the farmer should give attention to the texture of his soil before he worries about its richness. The conditions must first be made fit or comfortable for the growing of plants: then the stimulus of special or high feeding may be applied. But manures and fertilizers may aid in securing this good texture at the same time that they add plant-food. Yet fertilizer, however rich, may be applied to soils wholly without avail; and the best results from condensed or chemical fertilizers are usually obtained on soils which are in the best tilth. That is, it is almost useless to apply commercial fertilizers to lands which are not in proper physical condition for the best growth of crops.

SUGGESTIONS ON CHAPTER II

49*a*. The following extracts from Bulletin 119 of the Cornell Experiment Station illustrate the subject under discussion: "The other day, I secured one sample of soil from a very hard clay knoll upon which beans had been planted, but in which they were almost unable to germinate; another sample from a

Fig. 7. Examples of poor and good texture.

contiguous soil, in which beans were growing luxuriantly; and, as a third sample, I chipped a piece of rock off my house, which is built of stone of the neighborhood. All of these samples were taken to the chemist for analysis. The samples of soil which were actually taken to the chemist are shown in Fig. 7. The rock (sample III), was hard native stone."

The figures give the percentages of some of the leading constituents in the three materials.

	Moisture	*Nitrogen*	*Phosphoric acid*	*Potash*	*Lime*	*Organic matter*
I. Unproductive clay...	13.25	.08	.20	1.1	.41	3.19
II. Good bean land.......	15.95	.11	.17	.75	.61	5.45
III. Rock...............	......	......	.08	2.12	2.55	......

"In other words, the chemist says that the poorer soil—the one upon which I cannot grow beans—is the richer in mineral plant-food, and that the rock contains a most abundant supply of potash and about half as much phosphoric acid as the good bean soil.

"All this, after all, is not surprising, when we come to think of it. Every good farmer knows that a hard and lumpy soil will not grow good crops, no matter how much plant-food it may contain. A clay soil which has been producing good crops for any number of years may be so seriously injured by one injudicious plowing in a wet time as to ruin it for the growing of crops for two or three years. The injury lies in the modification of its physical texture, not in the lessening of its plant-food. A sandy soil may also be seriously impaired for the growing of any crop if the humus, or decaying organic matter, is allowed to burn out of it. It then becomes leachy, it quickly loses its moisture, and it becomes excessively hot in bright, sunny weather. Similar remarks may be applied to all soils. That is, *the texture or physical condition of the soil is nearly always more important than its mere richness in plant-food.*

"The first step in the enrichment of unproductive land is to improve its physical condition by means of careful and thorough tillage, by the addition of humus, and, perhaps, by under-drainage. It must first be put in such condition that plants can grow in it. After that, the addition of chemical fertilizers may pay by giving additional or redundant growth."

53*a*. Read Chapter ii. in King's "Soil." The following is quoted from that work, p. 72: "Suppose we take a marble exactly one inch in diameter. It will just slip inside a cube one inch on a side, and will hold a film of water 3.1416 square inches in area. But reduce the diameters of the marbles to one-tenth of an inch, and at least 1,000 of them will be required to fill the cubic inch, and their aggregate surface area will be 31.416 square inches. If, however, the diameters of these spheres be reduced to one-hundredth of an inch, 1,000,000 of them will be required to make a cubic inch, and their total surface area will then be 314.16 square inches. Suppose, again, the soil particles to have a diameter of one-thousandth of an inch. It will then require 1,000,000,000 of them to completely fill the cubic inch, while their aggregate surface area must measure 3141.59 square inches."

53*b*. Another illustration may be taken ("Texture of Soil and Conservation of Moisture," being a first lesson in the Cornell farmer's reading course): "Let us suppose the soil in one of your plowed fields is in little lumps of the uniform size of inch cubes—that is, one square inch on each side of the cube. How many square inches of surface has that cube exposed to root contact and moisture film? Now imagine that one of these inch cubes is broken up into smaller cubes measuring one-eighth of an inch,—how many square inches of surface will you now have exposed to root contact and film moisture? Now reflect what you have done in breaking up the inch cube of earth. The amount of earth has not been increased one atom; yet, by fining it, you have increased just eight times the root pasturage and surface for water film. The practical point of this lesson is that by superior tillage you can expand one acre into eight, or by neglectful management eight acres can be reduced to one. It also demonstrates why a skillful farmer can produce as much from fifty acres as a careless one can from four hundred, and also confirms the assertion that success in modern agriculture depends more on the size of the farmer than upon the size of the farm."

53*c*. This fining or dividing of the soil, therefore, increases the feeding area for roots; or, as Jethro Tull said, it extends the "root pasturage." "The value of simple tillage or fining of the land as a means of increasing its productivity was first clearly set forth in 1733 by Jethro Tull, in his 'New Horse Hoeing Husbandry.' The premises upon which Tull founded his system are erroneous. He supposed that plant roots actually take in or absorb the fine particles of the earth, and, therefore, the finer and more numerous these particles the more luxuriantly the plant will grow. His system of tillage, however, was correct, and his experiments and writings have had a most profound influence. If only one book of all the thousands which have been written on agriculture and rural affairs were to be preserved to future generations, I should want that honor conferred upon Tull's 'Horse Hoeing Husbandry.' It marked the beginning of the modern application of scientific methods to agriculture, and promulgated a system of treatment of the land which, in its essential principles, is now accepted by every good farmer, and the appreciation of which must increase to the end of time."—*Bailey, Bull. 119, Cornell Exp. Sta.* Tull died in 1740.

57*a*. "The actual contour of the water-table in an underdrained field, where the lines of tile are placed at distances of 33 feet and 4 feet below the surface of the ground, is shown in Fig. 8, which gives the contours as they existed forty-eight hours

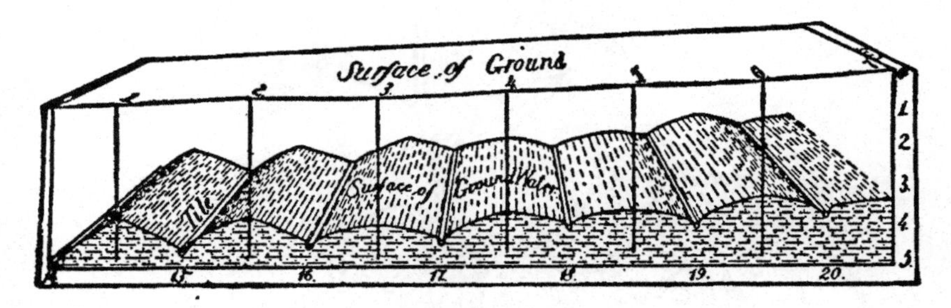

Fig. 8. Showing the actual contour of the water-table in a tile-drained field.

after a rainfall of .87 inches. In this case the height of the water midway between the lines of tile varied from 4 inches to 12 inches above the tops of the tile."—*King, The Soil, p. 259.*

58*a*. Read Roberts' "Fertility of the Land," pp. 303–312, on the physical effects of liming land; also "The Soil," p. 30, and Wheeler's "Liming of Soils," Farmers' Bulletin No. 77, U. S. Dept. Agric. The effects of lime in flocculating or mellowing clay may be observed by working up a ball of stiff clay with common water and a similar ball with lime water; the former will become hard on drying, but the latter will readily fall to pieces. Lime water may be made by shaking up a lump of lime in a bottle of water.

60*a*. One of the most forcible illustrations of the value of fine texture of soil is afforded by the result which the florist obtains in pots. He mixes and sifts his soils so that it is all amenable to root action, and he is able to raise a larger plant from a handful of soil than the general farmer grows from a half bushel. See Fig. 9.

Fig. 9. Showing the possibilities of a potful of soil.

CHAPTER III

THE MOISTURE IN THE SOIL

L. A. CLINTON

1. *Why Moisture Is Important*

61. However much plant-food there may be in the soil, plants cannot grow without the presence of water. This water is needed for two purposes: to dissolve the plant-food in the soil and thereby enable it to enter the plant; to contribute to the building of plant tissue and to the maintenance of the life of the plant.

62. A consideration of the amount of water required by plants in their growth shows why supplying plant-food alone does not insure the success of the crop. The amount of water used by some of the common crops in their development to maturity is approximately as follows:

Corn . . . . . 50 bus.	per acre	requires	1,500,000 lbs.	of water.
Potatoes . . . 200 bus.	"	"	1,268,000 lbs.	"
Oats . . . . . 29 bus.	"	"	1,192,000 lbs.	"

63. The failure of crops is more frequently due to lack of moisture than to any other one

cause. In certain sections of the country irrigation is successfully employed; but most farmers must depend upon the rainfall as the chief source for the supply of moisture.

2. *How Water Is Held in the Soil*

64. The water in the soil may be in one of three forms,—free, capillary, or hygroscopic water.

65. The free water of the soil is that which flows under the influence of gravity. It is this water which is removed in part by drains, and which is the source of supply for wells and springs. It is not utilized directly by cultivated plants, but it is valuable when removed a proper distance from the surface, because it serves as a reservoir from which moisture may be drawn by capillary action.

66. Capillary water is that which is held by adhesion to the soil particles, or in the interstices or openings between the particles. It is not controlled or influenced by gravity, but passes from one part of the soil to another, tending to keep the soil in equilibrium (or in uniform condition) so far as its moisture is concerned. The capillary water is the direct supply for plants, and it is this which should be most carefully provided for and saved.

67. Hygroscopic water is that which is held firmly as a film surrounding each particle of soil. It does not move under the influence of gravity or capillarity, and it is held so firmly that it is driven off only when the soil is exposed to a temperature of 212° Fahr. The dryest road-dust firmly holds its hygroscopic water, and it may constitute from 2 to 3 per cent or more of the weight of the soil. If of service to plants in any way, it is only during the most excessive droughts, in which case it may sustain the plants for a time, until capillary water is supplied.

68. Both capillary and hygroscopic water are frequently referred to as "film moisture," from the fact that they are held as a film of greater or less thickness around the soil particles. That part which has the most intimate and permanent contact with the particle is the hygroscopic water, and the outer part of the film, which may move away from the soil particle, is the capillary water. Very wet land is that which contains too much free water; whereas, soils which are dryish and crumbly usually contain sufficient water for the growing of plants. That is, lands in good condition for the growing of crops are moist, not wet; and we may, therefore, speak of the moisture of the soil rather than the water of the soil.

69. The free water of the soil is found at varying depths. Frequently it comes to the surface and oozes out as springs. Again it is many feet below the surface. The supply is maintained by rainfall, that part which is not held by capillary attraction or removed by surface drainage passing down to the level of the free water. In soils which are very porous and open, as gravelly soils, a large part of the rainfall passes down quickly, and such soils are said to be "leachy." With soils that are fine and compact and impervious, as in many clays, the water runs off by surface drainage, and not only is the supply of capillary water not increased to any perceptible degree, but the surface flowing removes valuable plant-food, causes erosion, and increases dangers from floods. Under these circumstances rainfall may be a detriment.

3. *How the Moisture-holding Capacity of the Soil May be Increased*

3a. *The capacity of the soil*

70. The first step toward utilizing the water of the soil is to so fit the land that the rainfall may be stored. In the winter months a large percentage of the rainfall is removed by surface

drainage, and in the summer months by evaporation. The soil should be put into such condition in the fall that it can readily absorb the winter rainfall. If the surface is hard, smooth and compacted, as is often the case with clay soils, it should be loosened with the plow and be left rough and uneven. If there is danger of surface erosion or washing, some quick-germinating seed (as rye or pea) may be sown in early fall. The plants prevent the rain from flowing away rapidly, and the roots bind the particles of soil in place.

71. The capacity of the soil to hold water depends upon its original constitution (whether clay, loam, sand, etc.) and upon the treatment which it has received. If the humus or decaying organic matter has been depleted, its moisture-holding capacity is diminished.

72. The capacity of the different soils to hold capillary and hygroscopic water (when dried at a temperature of 144°) is shown by the following-table:

Kind of soil	*Per cent (by weight) of moisture held in soil*	*Per cent (by volume) held in soil*	*Pounds of water in 1 cu. ft. of soil*
Silicious sand	25	37.9	27.3
Sandy clay	40	51.4	38.8
Loamy clay	50	57.3	41.4
Stiff brick-clay	61	62.9	45.4
Humus	181	69.8	50.1
Garden mold	89	67.3	48.4

3b. Capacity is increased by the addition of humus

73. A study of the above table reveals the fact that the humous soil (33) far exceeds any of the others in its ability to hold moisture. By long-continued cropping and tilling, without making proper returns in the way of green-manures or barn-manures, the humus may be so reduced that the soil consists very largely of mineral matter. One reason why newly cleared lands frequently give more satisfactory returns than lands which have been long cropped, is that the fresh land is rich in humus. The soil is consequently open and porous, and the rain which falls is quickly absorbed, and is largely retained as capillary or hygroscopic water.

74. The humus of the soil may be gradually increased by plowing under green-crops, by the use of barn-manures, by using cover-crops during the late summer and fall and plowing them under in the spring before they have used up the moisture which should be saved for the succeeding crop. These practices can be overdone, however, and the soil made so loose and open that the winds cause it to dry out quickly, and the power of drawing moisture from the stores of free water will be greatly lessened.

3c. Capacity may be increased by under-drainage

75. Drainage has an intimate relation to soil moisture. By drainage is meant the means employed for the removal of the surplus free water. Surface or open ditches may serve as conduits to carry off surface water, but as soil drains they are failures. The correct method for removing the surplus water of rainfall is to cause it to sink into the soil and be removed by under-drains. That which is removed by surface flow fails to impart any beneficial effect to the soil (69).

76. Lands which are well under-drained are porous. The rain which falls upon them passes down quickly, and is not removed by surface flow. It is removed only when the level of the free water rises to the level of the drain. By observing the action of drains which are of different depths, it has been found that after a protracted drought the drains which begin to flow first are those which are at the greatest depth, showing that as the level of the free water rises to the drain the flow begins, and that it is not removed to any considerable extent in its downward passage.

77. The sinking of the water through the soil does more good than merely to supply moisture. In the spring the rain is warmer than the soil,

and in passing down it gives up some of its heat, and the soil temperature is thereby raised. In the summer the rain is the cooler, and the soil parts with some of its heat. On lands which have been thoroughly under-drained, crops are far better able to withstand drought than those on land which needs drainage.

78. Few cultivated plants can thrive with their roots in free water. When the free water is near the surface, it is injurious in several ways: it limits the feeding space; it makes the soil cold in spring; it occupies the space which should be filled with air; it causes plant-food to be locked up; it dilutes the plant-food in solution; it prevents the action of micro-organisms; it causes the rainfall to be carried off largely by surface drainage. Thorough under-drainage tends to remove all these unfavorable conditions. If there is no effective under-drainage, either by natural or artificial channels, the water must escape by surface evaporation.

3d. The capacity is increased by proper tillage

79. Tillage enables soils to hold moisture by two means: by increasing the depth of the soil in which the plants can grow (that is, by increasing the depth of the reservoir), and by increasing the capillary power of the soil. We

have already seen (57, 75–78) that draining increases the depth of the soil ; so does deep plowing. Capillarity is increased by finely dividing or pulverizing the soil.

80. Increasing the capillarity increases the moisture-holding capacity of soils in two ways: it enables the soil to actually hold more moisture per square inch ; it enables it to draw up moisture from the free water of the lower subsoil (65).

81. By the action of capillary attraction, moisture moves from one layer of soil to another (66), usually from the lower to the upper, to supply the place of that which has been used by plants, or which has been lost by evaporation. The rapidity of movement and the force with which it is held depend upon various conditions. A soil in which the particles are somewhat large, as in sandy or gravelly soils, may, if well compacted, show considerable rapidity of movement, but weak power to retain moisture. The finer the division of the soil particles the greater is the surface presented. In finely divided clay soils, the movement of capillary water is slow but the retaining power is great. Occasionally it happens that the particles are so fine that the spaces disappear, and there is produced a condition through which moisture and air cannot pass. This state of

affairs is produced when clay soils are "puddled." It is evident, therefore, that soils which are either very loose or exceedingly finely pulverized are not in the best condition for the holding of moisture; but the danger of over-pulverizing is very small.

4. *The Conservation of Moisture*

82. By conservation of moisture is meant the prevention of all unnecessary waste of the capillary water of the soil, either through weeds or by evaporation. It is the saving and utilizing of moisture. The object is to make the water which seeks to escape from the surface pass through the cultivated plants. Plants require that their food be in solution. The moisture of the soil contains plant-food in solution. If this moisture is permitted to escape from the surface by evaporation, it leaves the plant-food at the surface. This food cannot nourish plants, because it is out of the range of their feeding roots. If the escape of the moisture is through the plants, there is created a moisture current towards the roots, and the plant-food is carried where it can be used to advantage.

83. Moisture rapidly rises to the surface by capillarity, to replace that which has evapo-

rated or has been used by plants, if the soil is in proper physical condition. Measures should be adopted to prevent this moisture from being lost by evaporation. The most practical and effective method is by establishing and maintaining a surface mulch of soil. By frequent use of implements of tillage, which loosen the soil to a depth of two or three inches, this mulch may be preserved and the moisture saved. The drier and looser this mulch, the more effective it is. This dry and loose surface breaks the capillary connection between the air and the moist under-soil, and has the effect of interposing a foreign body between the atmosphere and the earth. A board or a blanket laid on the earth has the same effect, and the soil is moist beneath it. This soil-mulch should be renewed, or repaired, in the growing season, as often as it becomes hard or baked, by means of shallow tillage.

SUGGESTIONS ON CHAPTER III

62*a*. To show that growing plants are constantly giving off large quantities of water through their foliage, grow corn, beans or squashes in rich soil in a flower-pot. Over the soil in the pot should be placed a rubber or oiled cloth covering, so that no moisture can come from this source. Then over the plant place a glass bell-jar or a common fruit-jar, and notice how rapidly tne moisture collects on the interior of the jar (Fig. 10). This experiment may be conducted even better in the field.

63*a*. Irrigation is admissable only in arid countries,—those in which the rainfall is very deficient,—and for special high-value crops. It is not to be advised for general crops in the country east of the Mississippi, for the rainfall is generally sufficient, if it is carefully saved.

66*a*. Capillary action, or capillarity, is due to the attraction of matter for matter. Capillary attraction is that force which

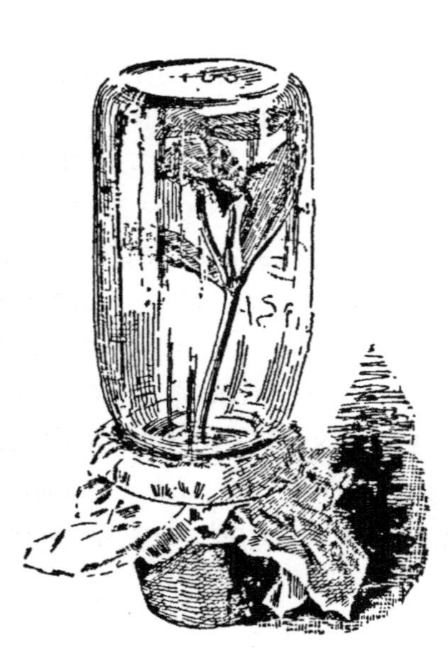

Fig. 10. How to show that plants give off moisture.

Fig. 11. To determine how much water a soil can hold.

causes a liquid to ascend or descend or move laterally through very small openings or tubes, or the interstices between fine particles of solid matter, or by which it is held to the surface of the particles themselves. The teacher should illustrate capillarity by the familiar experiment of standing tubes of glass in water. The smaller the bore of the tube, the higher the water rises. The oii rises in the wick by means of capillarity. The principle may be illustrated by filling straight (or argand) lamp chimneys with compacted dry soil and standing them in a dish of water.

68*a*. Film moisture can be illustrated by dipping a márble into water and observing the skin or film of moisture adhering to all sides. The most satisfactory conditions of soil moisture exist when each soil grain is covered by a film of water. The character of film moisture is changed by the thickness of the film. The thicker the film, the less the tension to the body, until it becomes so thick as to separate from that body and become a drop of water ; and it is then subject to the law of gravitation, and can travel but in one direction—downward. While in a state of film moisture, it is amenable to the law of capillary attraction, and can move in any direction, which means that it goes towards the thinnest films. The readiness with which water films travel can be seen by dipping a piece of cube sugar into coffee and observing how quickly the liquid pervades the lump of sugar. That soil moisture may move with the same facility as the coffee does in the sugar, it is necessary to have the soil grains in proper touch one with another ;—not so far apart but that the water films can reach one to the other, not so close as to impede the progress of the films. The two extremes in soil can be seen in loose gravel and hard clay.

70*a*. By rainfall is meant precipitation,—the fall of water in any form, as in rain, snow and hail.

72*a*. That different soils vary in their capacity to hold moisture may be illustrated by the following experiment : Provide several flower-pots of the same size and shape. The various soils should be thoroughly dried in an oven. At least four kinds of soil should be tested: gravel, sand, clay, and garden loam. Place an equal weight of each soil in the pots. Suspend one of the pots from a common spring-scales (Fig. 11). Notice the number of pounds and ounces registered. Now slowly pour water upon the soil until it is thoroughly saturated. Cover with a piece of oiled cloth or oiled paper, and allow it to drain until no more water will flow from it. The water which drains from the pot is the free water. The difference in weight of the pot of soil before soaking, and after the drainage, shows the amount of water held by capillarity.

74*a*. The plowing under of green-crops sometimes gives unsatisfactory results. If a heavy growth is plowed under when the soil does not contain sufficient moisture to cause ready decomposition, this layer of foreign matter prevents the passage of the water from the subsoil to the surface soil (Fig. 12). The crop which is then planted must necessarily feed for some time in the surface soil, and in case of prolonged drought a partial or complete failure of the crop may result. Heavy growths of cover-crops, as well as coarse, strawy manures, should be plowed under when there is sufficient moisture in the soil to cause decomposition. In case it is necessary to plow them under when the soil is dry, a heavy roller will so compact the soil that capillarity will be in part restored and decomposition hastened.

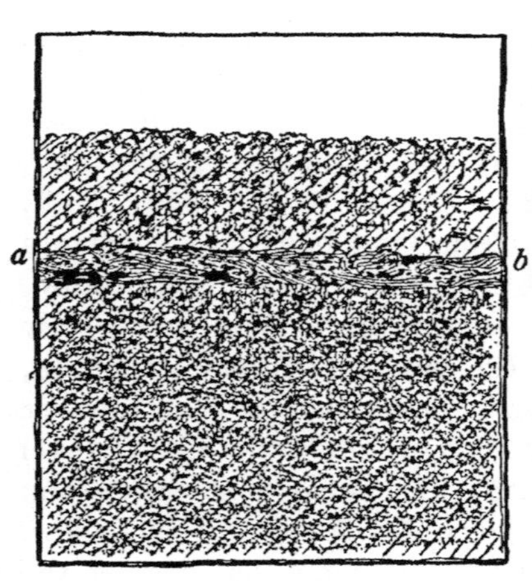

Fig. 12. The layer (*a b*) of undecomposed herbage.

75*a*. While surface drains are to be avoided, yet it frequently becomes necessary to provide a conduit or open ditch into which tile drains may open, or to remove flood water. It is a common error to have the banks too vertical. Through the action of frost or the tramping of stock, the banks are constantly requiring attention. The ditch should be wide, and the banks should have a gradual slope, as illustrated in Fig. 13. Grass-seed should be sown over the sides and bottom, so that the sod will prevent washing. One can drive across such a ditch. When possible, this ditch would be made the boundary of a field, or be placed near a fence.

76*a*. The depth at which tile drains should be placed must be determined by the nature of the soil. In very compact and impervious soils, as clay, the drains must be closer together and nearer the surface than in porous soils. Land may become so hard upon the surface that the water of rainfall never can pass down. By placing the drains shallow, the soil is rendered mellow and porous, water passes down readily, the level of free water is raised, and the surplus is removed.

76*b*. The distance apart at which drains should be placed is variable, but 30 feet is usually considered most advisable. The

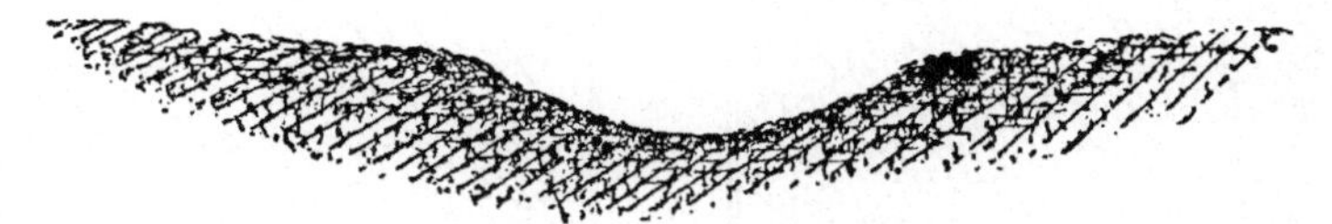

Fig. 13. Properly made open ditch.

level of the free water tends to rise higher at a point midway between drains, as shown in Fig. 8. If the drains are too far apart, this tendency may be greater than the tendency to move toward the drain. In soils through which the water moves somewhat readily, the drains may be farther removed than in close, impervious soils.

78*a*. In the spring, on undrained soils, free water remains for a considerable time near the surface; consequently the plant

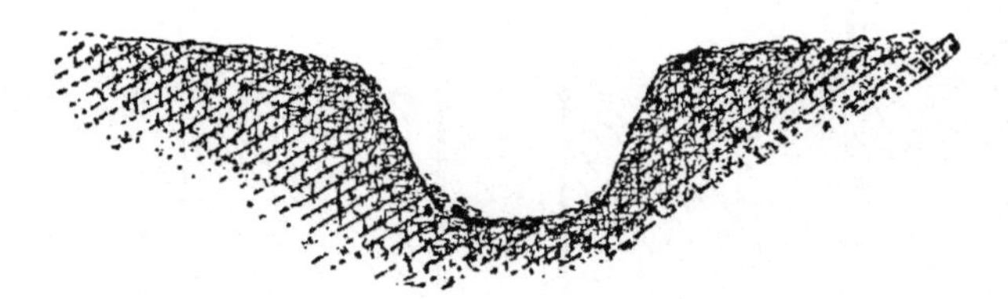

Fig. 14. Sides too steep.

roots cannot penetrate deeply into the soil. When the drought comes the surface is first affected, and the plants suffer at once. It is a well-known fact that tap-rooted plants are admirably fitted to withstand dry weather. Their feeders are deep in the soil. It is this condition which is obtained to a certain extent by under-drainage. The soil above the drain is made porous, the water which cannot be held by capillarity is quickly removed, the air penetrates, the soil becomes warm and congenial. Thus

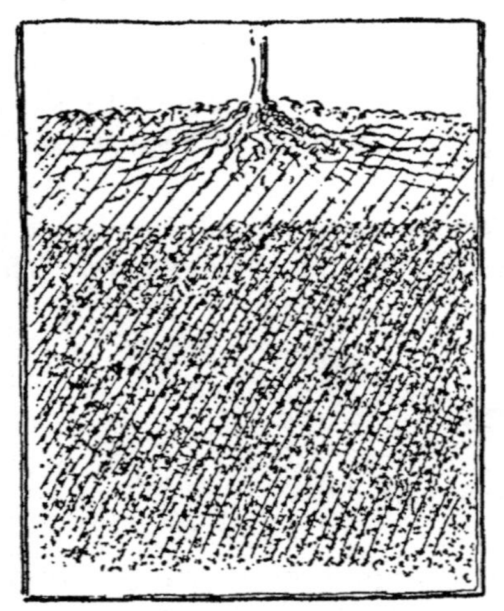

Fig. 15. Showing the condition which prevails in spring on cold, undrained soils,—when the water-table is too high.

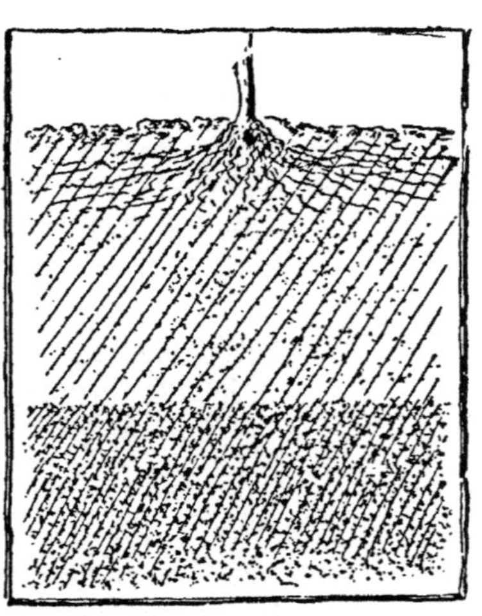

Fig. 16. When the drought comes, the plant is still shallow-rooted, and it suffers.

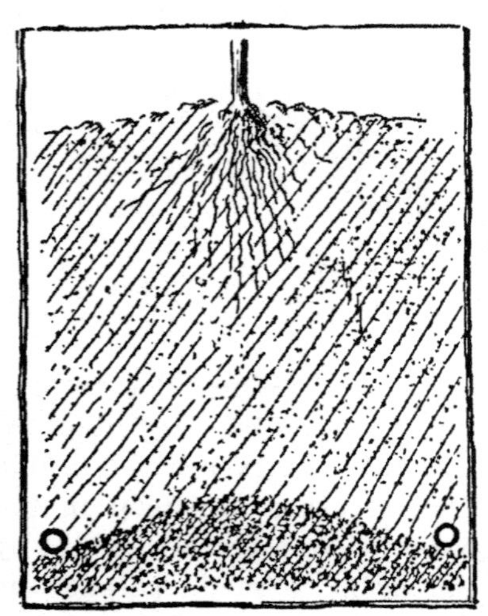

Fig. 17. On well-drained soils, the roots strike downwards.

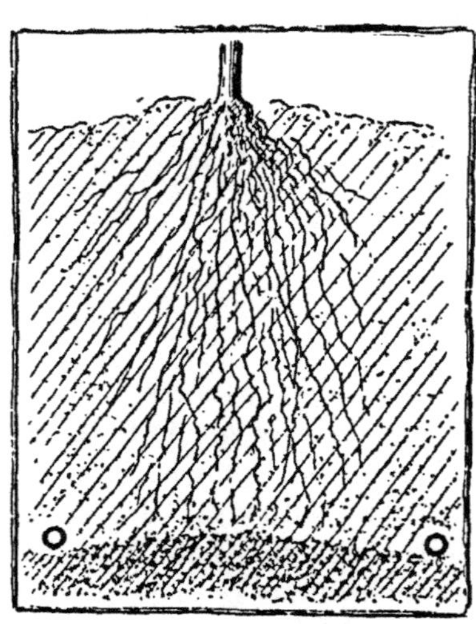

Fig 18. When the drought comes, the plant does not suffer.

plants are enabled early in their growth to send their roots down, and when drought comes they are not seriously injured. Figs. 15–18 illustrate this.

79*a*. The soil reservoir may be understood by likening it to a pan. A two-inch rainfall fills an inch-deep pan and runs it over; but if the depth is increased to two inches, none of the rain escapes. The hard-pan or water-table is the bottom of the soil reservoir. If this bottom is within a few inches of the surface, the ordinary rainfalls fill the soil so full that it is muddy, and some of the water may be lost by surface washing. Deep plowing lowers the bottom of the reservoir, and the soil holds more water and yet remains drier.

81*a*. Tillage operations should vary according to the nature of the soil. Those soils which are loose and porous should be compacted after plowing, so that the capillary connection may be restored between the surface and the subsoil. The roller may be used. With finely divided soils, which have a tendency to become too compact, only so much tillage should be given as is necessary to produce the proper degree of pulverization. It is possible to so compact and fine some soils, as clays, that the spaces between the soil particles is filled, and a condition is produced which prevents the rise of moisture by capillarity, and also prevents the absorption of rainfall and the passage of air.

81*b*. Of general farm crops, about three hundred pounds of water is used in the production of one pound of dry matter. An inch of rainfall weighs, approximately, one hundred and thirteen and one-half tons to the acre. The student will discover that the rainfall of the growing months may not be sufficient to supply the crop; hence the necessity of saving the rainfall of winter and spring.

83*a*. On the general subject of soil moisture and its conservation, read Chaps. v. and vi. in King's "Soil," and Chap. iv. in Roberts' "Fertility of the Land." Also consult Bull. 120, Cornell Exp. Sta.; Bull. 21, California Exp. Sta.; Bull. 43, Nebraska Exp. Sta., and Bull. 68, Kansas Exp. Sta.; publications of Division of Soils, U. S. Dept. Agriculture.

CHAPTER IV

THE TILLAGE OF THE SOIL

1. *What Tillage Is*

84. We have found (52, 79) that tillage is one of the means of improving the physical condition of the soil. By tillage is meant the stirring of the soil for the purpose of facilitating the growth of plants.

85. We may divide tillage into two general kinds,—tillage which covers the entire ground, and tillage which covers only that part of the ground which lies between the plants. The former we may call open or general tillage, and the latter inter-tillage. We practice open tillage before the seed is sown: it therefore prepares the land for the crop. We practice inter-tillage in fruit plantations and between the rows of crops: it therefore maintains the condition of the soil.

86. We may also speak of tillage as deep or shallow. In a general way, tillage is deep when it extends more than six inches into the ground. We also speak of surface tillage, when the

stirring is confined to the one, two or three uppermost inches of the soil.

2. *What Tillage Does*

87. Tillage improves the physical condition of the soil: by fining the soil and extending the feeding area for roots (53); by increasing the depth of the soil, or loosening it, so that plants obtain a deeper root-hold; by causing the soil to dry out and warm up in spring; by making the conditions of moisture and temperature more uniform throughout the growing season.

88. It aids in the saving of moisture: by increasing the water-holding capacity of the soil, or deepening the reservoir (79); by checking the evaporation (or conserving, or saving, moisture) by means of the surface-mulch (83). The former is the result of deep tillage, as deep plowing, and the latter of surface tillage.

89. It hastens and augments chemical action in the soil: by aiding to set free plant-food; by promoting nitrification (Chap. vi.); by admitting air to the soil; by lessening extremes of temperature; by hastening the decomposition of organic matter, as of green-crops or stable manures which are plowed under; by extending all these benefits to greater depths in the soil. In a very important sense, tillage is manure.

3. *How Tillage Is Performed*

3a. By deep-working tools

90. Plowing. We plow (*a*) to get the land in fit condition for planting, (*b*) to pulverize the soil, (*c*) to turn under manures, green-crops, and trash, (*d*) to deepen the soil, and thereby increase its storage capacity for water and extend the root pasturage, (*e*) to break up or to form a hard-pan, (*f*) to warm and dry the land, (*g*) to allow the weather to act on the soil. Passing over the first subject (*a*), we may explain the remaining objects of plowing.

91. (*b*) Plowing is the most efficient means of pulverizing the soil. That is, it is not enough that the soil be inverted: it must be ground and broken. For purposes of pulverization, the shape of the plow should be such as to twist the furrow-slice, causing it to break and crumble as it falls. The moldboard, therefore, should have a sharp, bold outward curve at its upper extremity; and the furrow-slice should be left in an inclined, or even nearly perpendicular position, rather than turned over flat.

92. (*c*) Since it is important that organic matter, as manures, shall quickly decay when turned under, the plowing should be done when the season is moist, as in early spring or in fall.

Clover and rye are also apt to become too hard and dry if allowed to grow to maturity. Herbage which does not decay quickly when plowed down may seriously injure the crop for that season (74*a*). For the covering of herbage, the furrow should be broad and deep; and if the land is to be surface-tilled shortly after the plowing, care should be taken that the furrow-slice turns down rather flat, so as to completely cover the plants.

93. (*d*) The deeper the plowing, the greater the water-storage reservoir will be, other things being equal; but the plowing may be so very deep as to bring the unproductive subsoil to the surface, in which case the increase of storage capacity may be overbalanced by the loss of available fertility. On most soils and for most crops, eight or nine inches is a sufficient depth for the plow. Shallow soils are both too dry and too wet. They are too dry, because much of the rainfall is lost in surface drainage or by very rapid evaporation. They are too wet after every hard rain, because the water is held near the surface (79*a*).

94. (*e*) If a hard-pan is near the surface, deep plowing will break it up, although the most permanent remedy may be under-drainage. In very porous soils, however, it may be necessary to form a hard-pan in order to prevent

leaching. This is done by plowing at the same depth each year, so that the land becomes compacted under the furrow. Loose and sandy lands may need shallow plowing rather than deep plowing.

95. (*f*) Land which is turned up loose soon dries out, because so much surface is exposed to the air. In spring, it is often necessary to make lands warm and dry, especially if such crops as corn and potatoes and cotton are to be planted; and this is done by very early plowing. The slices should not be turned down flat, but allowed to lie up loose and broken, and the harrow should not be used until the soil begins to be dry and crumbly. Care should be taken not to plow clay lands when wet, however, else they become lumpy and unmanageable.

96. (*g*) Freezing and thawing often pulverize and improve heavy lands, particularly clays. Fall plowing, therefore, may be advisable on lands which tend to remain lumpy. The results are best when the furrow-slices are left in a perpendicular position (as in Fig. 21), and when the harrow is not used until the following spring. Heavy clays tend to puddle (81) or to cement together if fall plowed, but the danger is least when there is herbage (as heavy sod or stubble) or manure on the land before it is plowed.

97. Subsoiling. When it is desired to loosen

or pulverize the land to a great depth, the subsoil plow is run in the furrow behind the ordinary plow. Subsoiling provides a deeper bed for roots, breaks up the hard-pan, and dries the soil. More permanent results are usually obtained by thorough under-drainage.

3b. By surface-working tools

98. Tillage by means of surface-working tools—as hoes, rakes, cultivators, harrows, clod-crushers—has the following objects: (*a*) to make a bed in which seeds can be sown or plants set, (*b*) to cover the seeds, (*c*) to pulverize the ground, (*d*) to establish and maintain an earth-mulch, (*e*) to destroy weeds. Aside from these specific benefits, surface tillage contributes to the general betterment of soil conditions, as outlined in 87, 88, 89.

99. In making the earth-mulch (the importance of which as a saver of moisture is fully explained in 82, 83), the other objects of surface tillage are also secured; therefore we may confine our attention to the earth-mulch for the present. The mulch is made by shallow tillage—about three inches deep, in field conditions—before the seeds are sown. The first tillage after plowing is usually with a heavy and coarse tool,—as a clod-crusher, cutaway harrow,

or spring-tooth harrow,—and its object is pulverization of the ground. The finishing is done with a small-toothed and lighter harrow; and this finishing provides the seed-bed and the soil-mulch.

100. The earth-mulch is destroyed by rains: the ground becomes baked. But even in dry times it becomes compact, and capillarity is restored between the under-soil and the air. Therefore, the mulch must be maintained or repaired. That is, the harrow or cultivator must be used as often as the ground becomes hard, particularly after every rain. In dry times, this surface tillage should usually be repeated every ten days,—oftener or less often as the judgment of the farmer may dictate. The drier the time and the country, the greater the necessity for maintaining the soil-mulch; but the mulch is of comparatively little effect in a dry time if the soil moisture was allowed to evaporate earlier in the season.

101. Surface tillage is usually looked upon only as a means of killing weeds, but we now see that we should till for tillage's sake,—to make the land more productive. If tillage is frequent and thorough—if the soil-mulch is maintained—weeds cannot obtain a start; and this is the ideal and profitable condition, to which, however, there may be exceptions.

3c. By compacting tools

102. The compacting tools are rollers, and the implements known as plankers or floats. The objects of rolling are: (*a*) to crush clods, (*b*) to smoothen the ground for the seed-bed, (*c*) to hasten germination of seeds, (*d*) to compact and solidify soils which are otherwise too loose and open, (*e*) to put the land in such condition that other tools can act efficiently, (*f*) to facilitate the marking-out of land.

103. By compacting the surface soil, the roller re-establishes the capillary connection between the under-soil and the air: that is, it destroys the earth-mulch. In its passage upwards, the soil moisture supplies the seeds with water; and the particles of the soil are in intimate contact with the seeds, and, therefore, with the soil moisture. If the surface of rolled lands is moister than loose-tilled lands, therefore, it is because the moisture is passing off into the air and is being lost.

104. The rolling of lands, then, sacrifices soil moisture. The rolled or compacted surface should not be allowed to remain, but the earth-mulch should be quickly restored, to prevent evaporation, particularly in dry weather. When the object of rolling is to hasten germination, however, the surface cannot be tilled at once;

but if the seed is in rows or hills, as maize or garden vegetables, tillage should begin as soon as the plants have appeared.

SUGGESTIONS ON CHAPTER IV

84*a*. Tillage is a specific or special word, and is much better than the more general word *culture*, when one is speaking of the stirring of the soil. The culture of a crop properly comprises tillage, pruning, fertilizing, and other good care.

85*a*. For the origin of the word *inter-tillage*, see foot-note in Roberts' "Fertility of the Land," p. 69.

88*a*. It should be observed that surface tillage saves moisture by preventing evaporation, not, as commonly supposed, by causing the soil to absorb moisture from the atmosphere. When moisture is most needed, is the season in which the air is dryer than the soil.

89*a*. To illustrate the importance of air, select a thrifty plant, other than aquatic plant, growing in a florist's pot, and exclude all the air by keeping the soil saturated with water, or even by keeping the bottom of the plant standing deep in water, and note the checking of growth, and, in time, the decline of the plant. The remarks on draining (65, 78) show how undrained soils are often saturated with water; and no matter how much raw material for plant-food may exist in such a soil, it is unavailable to the plant. The reader can now guess why crops are poor and yellow on flat lands in wet seasons. On the importance of air in soils, read Chapter ix. of King's "Soil."

89*b*. On the effects and necessity of tillage, read Chapter iii. in Roberts' "Fertility of the Land," and Chapter xii. in King's "Soil." A most interesting diversion in this connection is a perusal of Jethro Tull's famous book on "Horse-Hoeing Husbandry" (53*c*). Copies of Cobbett's edition may frequently be found in antiquarian book stores.

91*a*. The trench left by the plow is a furrow. The earth

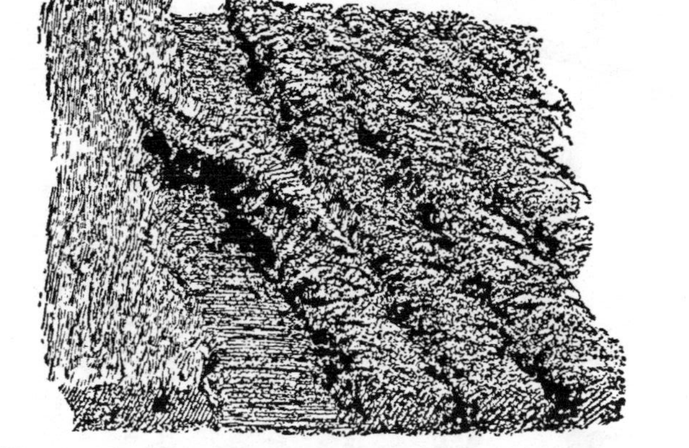

Fig 19. The earth completely inverted, making handsome but poor plowing.

Fig. 20. Furrow-slice standing on edge.

Fig. 21. Ideal plowing, the furrow-slice being broken or pulverized.

Fig. 22. The ideal general-purpose plow.

which is turned out of the furrow is a furrow-slice. In common speech, however, the word furrow is often used for the furrow-slice.

91*b*. The accompanying pictures, adapted from Roberts' "Fertility of the Land," illustrate different types of plow-work. Fig. 19 shows the furrow-slice completely inverted. This kind of plowing looks well, but it is not desirable unless the object is to bury weeds or a green-crop. The furrow-slices are not broken

Fig. 23. A subsoil plow. Fig. 24. A smoothing harrow.

and pulverized, and they are in such position that the harrow cannot tear them to pieces. Fig. 20 represents work which is better, for most conditions, although the slices are not pulverized. Fig. 21 shows ideal plowing.

91*c*. The ideal plow for general farm work, in Roberts' opinion, is shown in Fig. 22. Observe the "quick" or sharp curve of the moldboard. For an excellent sketch of the development of the plow, consult Chapter ii. of Roberts' "Fertility of the Land."

93*a*. About 12 to 20 per cent of moisture in the soil is the ideal condition for most plants. Let the pupil figure out what the percentage will be after a rainfall of one inch on soils that are four inches deep and eight inches deep. Consult Roberts, "Fertility of the Land," pp. 77 to 79.

94*a*. By hard-pan is meant very hard and more or less impervious subsoil. Some subsoils are loose; others are so hard as to prevent the downward movement of water and roots (79*a*).

Fig. 25. The loose mulch on forest soils.

Fig. 26. The soil-mulch on tilled lands.

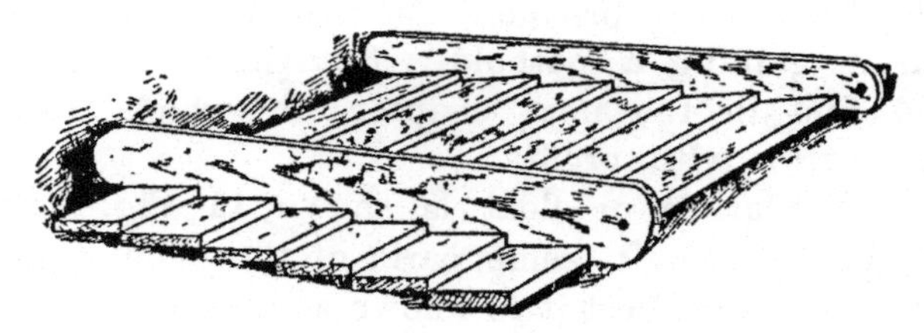

Fig. 27. A home-made planker.

Fig. 28. Showing the effect of the roller in compacting the surface layer.

Fig. 29. Showing how the soil-mulch should be restored by tillage after the roller has been used.

97*a*. The subsoil plow does not turn a furrow (Fig. 23). It is drawn by an extra team, which follows the ordinary plowing.

99*a*. A useful tool for making and maintaining the soil-mulch is the smoothing harrow shown in Fig. 24. On hard lands, however, heavier and more vigorous tools must be used.

99*b*. Observe how moist the soil is in forests, even in dry times. This condition is due partly to the forest shade, but perhaps chiefly to the mulch of leaves on the ground (Fig. 25).

101*a*. Some farmers are always asking how to kill weeds, as if this were the chief end of farming. But good farmers seldom worry about weeds, because that management of the farm which makes land the most productive is also the one which prevents weeds from gaining a foothold. But there are some cases, as we shall find in the next chapter, in which weeds may be allowed to grow with profit.

102*a*. A planker or float is shown in Fig. 27. This is a home-made device. In some parts of the country it is called a slicker; and in the West it is known as a drag. In the East, the word drag is synonymous with harrow.

104*a*. To determine when and how much to roll land, is one of the most difficult of agricultural operations. This is because the good effects are so often followed by the ill effects of loss of moisture and of puddling of hard lands when heavy rains follow. Whenever the object of rolling is to compact loose lands or merely to crush the clods, the work should be quickly followed by the harrow or cultivator. Compare Figs. 28 and 29. For fuller advice on rolling, consult Roberts, "Fertility of the Land," p. 102; L. A. Clinton, Bull. 120, Cornell Exp. Sta.; Bailey, "Principles of Fruit-Growing," p. 152.

Chapter V

ENRICHING THE SOIL—FARM RESOURCES

1. *What Farm Resources Are*

105. The real fertility of the land is its power to produce crops. It is sometimes said to be the richness of the soil in elements of plant-food; but soils with much plant-food may still be unproductive. Fertility is productive power. It is the result of good physical condition and an abundance of available plant-food.

106. We have found (in Chapters ii., iii. and iv.) that the first step towards increasing the productiveness of soil is to improve its physical texture. This improvement is accomplished both by mechanical means,—as tillage and drainage,—and by the addition of humus. But humus also adds plant-food, and, therefore, directly enriches the land.

107. We have seen (34) that humus is supplied, in practice, by cropping,—that is, by vegetable matter left on the ground after the crop is removed, or by crops plowed under;

and by stable manures and other direct applications.

2. *Cropping Resources*

2a. *The kinds of green-manures*

108. The stubbles of grain, clover, grass and sowed corn add considerable humus to the soil, and there is also much vegetable fiber left in the ground in the roots; and the refuse left from potatoes and garden crops is often important. Sometimes the stubble and roots are nearly as valuable for ameliorating the soil as the part which is removed from the land. This is especially true in clover, particularly if it is not cut close to the ground. Roberts reports that a second-growth of clover, two years from seeding, gave 5,417 pounds per acre of top and 2,368 pounds of roots in the upper eight inches of soil; and the roots usually extend to three or four times that depth.

109. Humus is often secured by growing crops for that particular purpose; that is, by the practice of green-manuring. Green-manure crops are of three categories: (*a*) regular or full-season crops, which occupy the land for one or more seasons before they are plowed under, or until they have reached nearly or quite their full growth; (*b*) catch-crops, which are grown in the seasons between other crops;

(*c*) cover-crops, which are sown late in the season for the purpose of protecting the soil during winter as well as for green-manuring.

110. Green-manuring crops may be again divided into those which gather nitrogen and those which do not,—or those which have the power of using the nitrogen (see Chapter vi.) of the air, and those which obtain all their nitrogen directly from the soil. The nitrogen-gatherers leave their nitrogen in the soil, when they decay, for the use of other plants. The nitrogen-gatherers are the leguminous plants, or those which belong to the pea family, as all kinds of peas and beans, clovers, alfalfa, vetch. The other class, or nitrogen-consumers, comprises all other plants used for green-manuring, as rye, oats, rape, mustard, buckwheat, maize.

111. In general, the best green-manure crops are the legumes,—red clover for the North, alfalfa for dry regions, cow-peas and Japan clover for the South. With the exception of the cow-peas, these crops require one or more seasons for full development, and, therefore, cannot be used in intensive farming.

2b. The management of green-manures

112. The ideal green-manuring is that which is a part of a regular rotation,—the green-

manure crop, or the stubble or sod, occurring regularly once every few years, in alternation with wheat, potatoes and other staple crops. This, however, is possible only with general or mixed husbandry (4*a*). In market-gardening, and other intensive farming, catch-crops are often used. In fruit-growing, cover-crops are frequently used.

113. But even in intensive farming, the land sometimes becomes unproductive from too continuous cropping with one thing, and the too persistent use of one kind of fertilizer. It is then often "rested" by seeding it to clover; but the good effects are not the result of a rest, but of rotation or change of crop.

114. It is necessary to distinguish between the effects of green-crops in improving soil texture and their effects in enriching the soil; for soils which may need improving in texture may not need enriching. In fruit-growing this is often true; and the heavy addition of nitrogen (which conduces to growth of wood) may cause the plants to grow too heavily and to bear little, and to be too susceptible to disease and to cold. In such cases, the nitrogen-consumers are the better crops. One must be careful not to induce an over-growth in grapes, peaches, apricots, and pears.

115. On hard and poor lands, it is often

difficult to secure a "catch" of clover. In such cases, it is well to begin with fall-sown rye or field peas. When the soil has become mellow, clover may be successful.

116. Cover-crops are used mostly in fruit plantations. They are sown in midsummer, or later, after tillage is completed,—for tillage should cease early, in order that the fruit plants will not grow too heavily and too late. The cover is plowed under early the following spring (74*a*). The cover checks the growth of the fruit plants, prevents the land from washing and puddling, holds the rainfall until it can soak into the soil, causes the soil to dry out early in spring, lessens injury from frost.

117. Weeds often make good cover-crops. The chief difficulty is that they cannot be relied upon to appear when and where and in the quantity wanted, and some kinds may be difficult to eradicate (101*a*).

3. *Direct Applications*

3*a*. *Stable manures*

118. The best direct application which the farmer can make to his land, from his home resources, is stable manure. It supplies both humus and plant-food.

119. The value of manure depends upon (*a*) the kind of animal from which it is made, (*b*) the feed which the animal receives, (*c*) the amount of bedding or litter which it contains, (*d*) the way in which it is kept or housed.

120. Some of the most valuable constituents of manure are soluble, and are, therefore, removed by water. Consequently, manures should be housed to protect them from rain. A covered barn-yard is the ideal place in which to keep manures, for they are not only protected from weather, but, if the manure contains enough straw or litter, it makes an agreeable bed upon which stock may tramp, and it absorbs the liquids; and if it is spread in the yard as it is made and well tramped by stock, its tendency to heat is reduced. In six months' exposure to weather, manures usually lose more than half of their available plant-food.

121. The more completely rotted the manure, the sooner does it become thoroughly incorporated with the soil; and the decay of the coarse parts renders their plant-food more available. If the rotting proceeds under cover or in a compost pile (34*a*, Fig. 5), there should be little loss of plant-food by leaching.

122. If manure cannot be sheltered, it should be spread on the fields as fast as

made. There is practically no loss of plant-food from evaporation, and the part which leaches is caught by the soil. Loose or strawy manure which lies too long on the ground, however, may become so dry that it does not quickly decay when plowed under; if applied very thick, it prevents heavy soils from drying out, and thereby delays spring work.

3b. Other dressings

123. Muck is often useful as a source of humus, but it generally contains little directly available plant-food. It is generally improved if dug and allowed to weather some time before it is put on the land. Dry muck is very useful in stables and covered barn-yards to absorb the liquids; and its value as a dressing for the land is thereby increased.

124. Peat, when decomposed and soil-like, becomes muck. Peat, therefore, is less valuable than muck as a dressing until it has been thoroughly broken up and decomposed by weathering or composting.

125. Marl is usually not rich in available plant-food, but, like muck, it may be valuable to improve the physical condition of the soil. But only in exceptional cases is it worth hauling great distances.

126. Such materials as sawdust, straw,

leaves, pomace, are generally more valuable for the improving of the texture of the soil than for the direct addition of plant-food. If the soil is loose, dry and leachy, or if it is very hard, compact and retentive, these materials may benefit it. To determine the value of such materials in plant-food, one must consult tables of their composition in books; and the more thoroughly they are rotted, the more available are their constituents.

SUGGESTIONS ON CHAPTER V

108*a*. "The proportion of roots to tops [in clovers] varies widely. The medium red clover, one year from seeding, gives a much larger proportion of roots to tops than clover two years from seeding. Red clover which produces two tons per acre may be expected to furnish potentially to the soil, after the first cutting, in roots and stubble, 40 to 60 pounds of nitrogen, 20 to 25 pounds of phosphoric acid, and 30 to 50 pounds of potash. Thirty bushels of wheat * * * and 2,700 pounds of straw, would remove approximately 46 pounds of nitrogen, 20 pounds of phosphoric acid, and 26 pounds of potash."—*Roberts*, "*Fertility of the Land*," *345*.

109*a*. Accessible discussions of green-manuring are to be found in Chap. xiv., "Fertility of the Land;" pp. 117–123, Voorhees' "Fertilizers." Cover-crops in relation to fruit-culture are discussed in pp. 184–202 of Bailey's "Principles of Fruit-Growing."

111*a*. Intensive farming is "high-culture" farming. It is farming on a comparatively small scale, when the land is kept constantly in productive crop, with the best of tillage, and the free use of manures and fertilizers. The land is forced to its

Fig. 30. A covered barn-yard, in which manure is saved and the stock protected.

Fig. 31. A common type of barn-yard. The stains on the barn show where the manure was baptized from the eaves; and the mud-puddle shows where much of the fertility has gone.

utmost capacity. Market-gardening and forcing-house culture are examples.

111*b*. Extensive farming is general husbandry, especially when done on a large scale and without forceful methods of tillage and cropping. Grain-farming and stock-raising are examples.

120*a*. A covered barn-yard is shown in Fig. 30. This is a basement under the farm barn at Cornell University. This affords a protected place in which the stock may exercise in cold weather; and if the cattle are dehorned, they remain to-

Fig. 32. A handy and economical stable, with cattle-racks, a manure trough (behind which is a walk), and a small shed at the rear, with a hollowed cement bottom, for the storage of the manure.

gether peaceably. Such an area not only saves the manure, but it adds to the welfare and value of the stock. Compare this with the commoner type of yard, as shown in Fig. 31. A handy and efficient arrangement for the saving of manure is shown in Fig. 32. For general discussions on farm manures and methods of saving and handling them, consult Roberts, "Fertility of the Land," Chapters vi., vii., viii., ix.

126*a*. Muck, marl, and other materials of this class are considered in Voorhees' "Fertilizers," Chapter vi., and in Roberts' "Fertility, Chapter xiii.;" and the appendix to the latter work has full tables of the fertilizer constituents of very many substances.

Chapter VI

ENRICHING THE SOIL—COMMERCIAL RESOURCES

G. W. CAVANAUGH

1. *The Elements in the Soil*

127. Chemically, a fertile soil is one containing an abundance of available plant-food. The substances which are necessary for the growth and welfare of plants are called plant-foods. There are thirteen essential elements of the plant-food. Nine of these are derived from the mineral part of the soil,—phosphorus, silicon, sulfur, chlorine, iron, calcium, magnesium, sodium and potassium. Nitrogen is contained in the humus. Water supplies the hydrogen and oxygen to the roots. Carbon comes from the air. Fortunately, the greater part of the plant-food elements of the soil always exist in quantities more than sufficient to supply any possible need of the plants.

128. Three of these elements are often deficient in the soil; or, if present, they may not

be in condition to be used by the plant. These are nitrogen, phosphorus, and potassium. A fourth plant-food is also sometimes deficient,—calcium. These four substances, therefore, are the ones which the farmer needs to consider when fertilizing the land.

129. Before the plant can use any of these elements of plant-food in the soil, they must become dissolved in the soil water, which is absorbed by roots.

130. While all plants need certain elements for their growth, they cannot use the elements in their elemental or uncombined forms. In fact, the elements as such do not exist in the soil. They are united with each other in compounds, and it is by absorbing the compounds that the plants obtain the necessary elements. Phosphorus is essential to the life of plants, but it is never used by them in the form of elemental phosphorus. It is always in some compound, as phosphoric acid or a phosphate.

131. When the compounds exist in such condition as to be readily absorbed by the roots, the soil is said to contain available plant-food. Often there is sufficient plant-food present, but not in condition to be taken up by the plants. It is then said to be unavailable, or to be locked up. Availability is determined by two factors: by the substance being

soluble in soil water; by its being of such composition that the plant will use it.

132. One problem for the agriculturist is to secure available plant-food, and to determine whether it is better to unlock the plant-food in the soil by means of tillage, or to supply the elements in some manure or fertilizer.

133. Barn manures are not always to be had, and they are variable in composition. It is often advisable, therefore, to substitute commercial or concentrated fertilizers, in which the constituents are of known amounts and often readily available. Barn manures are bulky. Even manure of cattle from a covered yard contains as high as 70 or 75 per cent of water, and usually less than 1 per cent of nitrogen, phosphoric acid or potash. If it were not for its influence in improving the physical effects of the soil, stable manure would have comparatively little value.

2. *Nitrogen*

134. Nitrogen is the most important element which the farmer adds to his soil. It comprises part of all green and woody parts of plants. It seems to be the element most intimately associated with rapid growth in plants. Plants that feed excessively on nitrogen tend to pro-

duce large leaves and stalks, while the hardiness may suffer. On the other hand, insufficient nitrogen is almost certain to result in dwarfing and loss of vitality. It must receive attention, also, because it tends to leach from the soil.

135. In a pure or elemental state, nitrogen is an invisible gas. It comprises four-fifths of the atmosphere. And yet, with this vast amount about us, it is the most expensive element of plant-food. The nitrogen of the air can not be used by the great majority of plants, because it is in what is known as a free or uncombined state. The sources of nitrogen for plants are ammonia, nitrates, or in some composition formed by animals or plants (that is, in some organic form).

136. If the gas nitrogen be combined with the gas hydrogen, there will be formed ammonia ($N H_3$). From this the plants can derive, indirectly, their supply of nitrogen. Another compound of nitrogen is called nitric acid, which is composed of nitrogen, hydrogen, and oxygen ($H N O_3$). When some mineral element takes the place of the hydrogen in this combination, the compound is called a nitrate: as $Na N O_3$, nitrate of soda; $K N O_3$, nitrate of potash, or saltpetre. Both ammonia and nitrates are found in the soil in small

quantities, but only in a fertile soil in sufficient amounts to supply the plant with nitrogen.

137. Humus is the great storehouse of nitrogen. Humus does not dissolve in water, and so serves as a means of retaining the nitrogen against leaching. But if the nitrogen remained always in the humus, it would not be available to plants, since to be absorbed it must dissolve in the soil-water. Fortunately there is a process whereby the nitrogen in the insoluble humus is made to be available. This process is the work of germs or micro-organisms (35, 35*a*). These germs are of several kinds. One kind works upon the humus and changes its nitrogen into ammonia, and other kinds change the ammonia into nitric acid. This process of changing nitrogen into the form of nitric acid or nitrate is called nitrification. It is probable that nitrogen enters the plant chiefly in form of nitrate, so that all other forms of nitrogen must undergo nitrification, or be nitrified, before they are of use. Since tillage promotes the activities of the micro-organisms (35, 52, 89), it thereby increases the supply of available nitrogen.

138. It has been stated (135) that the great quantity of nitrogen in the atmosphere is not available to most plants, because it is not in a combined state. There are certain plants,

however, which have the power of drawing upon this supply for their nitrogen. They are the leguminous plants, and include the clovers, peas and beans (110). These plants have knobs or nodules growing upon their roots. These nodules are the homes of germs; and these germs seize upon the nitrogen of the air and turn it over to the plant. This process is known as the fixation of nitrogen. Then if these crops are plowed under they not only add humus from their vegetable substance, but nitrogen which has been gathered from the air.

139. The nitrogen added in green-crops or humus must go through the process of nitrification before it is available to the plant. Sometimes this process does not furnish nitric acid fast enough to supply rapidly growing plants, and then a form of available nitrogen may be added direct. This can be done by using nitrate of soda or sulfate of ammonia. The former is mined in Chile; the latter is a substance obtained from gas works, where the ammonia formed from the nitrogen that was in the coal or wood is caught in sulfuric acid or oil of vitriol. These two substances, together with dried blood from the slaughter houses, constitute the best sources of nitrogen in commercial fertilizers.

3. *Phosphoric Acid*

140. Phosphoric acid is, next to nitrogen, the most important plant-food to be applied to land, and of the mineral constituents it is the most important. It is a constituent of all soils, though the amount may be variable. It is particularly needed to insure hardiness and fruitfulness. Consequently the different grain crops are large users of phosphoric acid. A liberal supply of available phosphoric acid is necessary to young plants to give them strength and vigor.

141. As humus decays or decomposes in the soil it not only supplies nitrogen, but it also makes some of the phosphoric acid available. Hence when the humus diminishes in the soil, there is often a corresponding lack of available phosphoric acid. Barn manures contain a considerable quantity of phosphoric acid. Soils which contain a fair supply of humus do not necessarily have enough of phosphoric acid. To such soils phosphoric acid may be supplied in an available form in acid phosphates.

142. Pure phosphoric acid (P_2O_5), however, is not used directly as a plant-food, but only when it is combined with some other substance, as lime. One of the chief sources of phosphoric acid is bone, in which it is found combined

with lime. The animals obtained the phosphoric acid from the plants they ate, which in their turn secured it from the soil. Another great source are the deposits of phosphatic rocks in the Carolinas, Florida and Canada. In these rocks the phosphoric acid and lime are combined in the same way as in bones.

143. Bones and phosphoric rocks do not dissolve in water, and consequently the phosphoric acid they contain is not easily absorbed by roots. These materials, therefore, are commonly treated with acid, to make the phosphoric acid soluble; and the material is then known as an acid phosphate.

144. In bones, one part of phosphoric acid (P_2O_5) is combined with three parts of lime (CaO), and can be expressed as follows:

$$\left.\begin{array}{l}\text{Lime}\\ \text{Lime}\\ \text{Lime}\end{array}\right\}\text{Phosphoric acid; or, }\left.\begin{array}{l}CaO\\ CaO\\ CaO\end{array}\right\}P_2O_5$$

This substance is tri- (or three) calcic phosphate, and is insoluble. When sulfuric acid (or oil of vitriol) and water are brought in contact with the bones, part of the lime leaves the phosphoric acid, and its place is taken by water. If one part of the lime is united with the sulfuric acid, then there results a substance which can be written thus:

$$\left.\begin{array}{l}\text{Water}\\ \text{Lime}\\ \text{Lime}\end{array}\right\}\text{Phosphoric acid; or, }\left.\begin{array}{r}H_2O\\ CaO\\ CaO\end{array}\right\}P_2O_5$$

This is di- (or two) calcic phosphate. This is insoluble in rain-water, but is readily dissolved and used by roots.

145. If two parts of the lime be united with sulfuric acid and their places be taken by water, there remains:

$$\left.\begin{array}{l}\text{Water}\\ \text{Water}\\ \text{Lime}\end{array}\right\}\text{Phosphoric acid; or, }\left.\begin{array}{r}H_2O\\ H_2O\\ CaO\end{array}\right\}P_2O_5$$

This is mono- (or one) calcic phosphate. This is readily soluble in soil water, but in the soil it tends to become insoluble, or to revert to the dicalcic form (and is then said to be "reverted"), and some of it may eventually become tricalcic and unavailable. The lime that is removed by the sulfuric acid unites with the sulfuric acid to form calcium sulfate; that is, plaster or gypsum ($CaSO_4$). The dicalcic and monocalcic are the forms that are known as acid phosphate, and sold in commercial fertilizers.

4. *Potash* (*potassium oxide*, K_2O)

146. Next to phosphoric acid, potash is the most important mineral plant-food. It is placed after phosphoric acid in importance not be-

cause plants can better do without it, but because it is usually more abundant in soils. Potash has an important office in the production of firm, woody tissue and of starch, and it is thought to be particularly needed by fruit-plants, potatoes, and root crops. It is generally deficient in sandy and peaty soils.

147. Like phosphoric acid, potash becomes available with a liberal supply of humus and by good tillage; and the potash in barn manures is soluble and valuable. Whenever wood ashes can be cheaply obtained they form a valuable source of potash, for the potash taken from the soil by the trees remains in the ashes when the wood is burned.

148. Potash is found in great deposits in Germany, very much as common salt is found in the United States. There it is mined and sold. It can be bought in the form known as the muriate of potash, or more properly potassium chlorid, KCl. Another form of potash is the sulfate, K_2SO_4. The sulfate costs a little more than the other, because it is made from the muriate. For general purposes, the muriate is recommended over the sulfate because it is cheaper; but the muriate has a deleterious effect on tobacco, and it is thought to give less satisfactory results on sugar-cane and potatoes.

5. *Amendments*

149. Substances which contain only traces of the important or available plant-foods often have a beneficial effect on soil. Lime and salt are examples. Though they may not add to the soil any needed plant-food, the plants are enabled by their presence to utilize more of the plant-food already in the soil. Such materials are known as amendments (58).

150. It is often difficult to decide, in any particular case, just how an amendment produces its effect. It may be that the mechanical condition of the soil is improved, its water-holding capacity increased, its acidity or sourness neutralized, or its plant-food unlocked.

151. Lime. Soils sometimes become sour, and may then be uncomfortable for some plants. One of the reasons why plants do not thrive well in sour soils is that it is difficult to obtain sufficient nitrogen in the form of nitrates. The germs which carry on the process of nitrification are unable to do their work in sour soils. The soil acid can be neutralized — the soil sweetened — by applying lime (which is calcium oxide, CaO).

152. Lime may be applied in the form of water-slaked lime, such as is obtained by adding water to quick-lime till it crumbles, or by air-slaked lime. Quick-lime usually gives the better

results, particularly when it is desired to improve the texture of clay soils (58, 58*a*).

153. A soil may be tested to determine if it is acid by placing a piece of blue litmus paper (kept at drug stores) against the moist soil. If the paper reddens and remains so after drying, it shows the presence of an acid in the soil. It is best to apply the paper not to the top of the soil, but to the side of a hole such as would be made by inserting a spade and moving it to and fro.

6. *Commercial Fertilizers*

6*a*. *What they are*

154. Under the name of commercial fertilizers, one can buy the various forms of nitrogen, phosphoric acid and potash. These elements may be purchased singly or mixed in any combination. A fertilizer containing all three is called a complete manure or fertilizer. In buying, one should be guided by the guaranteed analysis and not by any particular name or brand.

155. The commercial value of nitrogen is about three times that of either phosphoric acid or potash, which are approximately 5 cents per pound. The prices of these elements may vary, but the following will serve as an illustration of the computing of relative values of different fer-

tilizers (remembering that 1 per cent means one pound in a hundred, or twenty pounds in a ton):

No. 1. GUARANTEED ANALYSIS

Nitrogen	1.60 to 2.00	per cent
Phosphoric acid available .	7.00 to 8.00	" "
Potash	2.00 to 3.50	" "

Cost per ton, $29.

Multiplying the lowest figure representing the per cent of the given element by 20, and calculating the value from the price per pound, we have in No. 1:

Nitrogen . . .	1.60 × 20 =	32 lbs.	@15c. =	$4 80
Phosphoric acid	7 × 20 =	140 lbs.	@ 5c. =	7 00
Potash	2 × 20 =	40 lbs.	@ 5c. =	2 00
Commercial value per ton				$13 80

156. Another example of computation may be taken:

No. 2. GUARANTEED ANALYSIS

Nitrogen	3.30 to 4.00	per cent
Phosphoric acid available .	8.00 to 10.00	" "
Potash	7.00 to 8.00	" "

Cost per ton, $38.

Its value is calculated the same as No. 1:

Nitrogen . . .	3.30 × 20 =	66 lbs.	@15c. =	$9 90
Phosphoric acid	8.00 × 20 =	160 lbs.	@ 5c. =	8 00
Potash	7.00 × 20 =	140 lbs.	@ 5c. =	7 00
Commercial value				$24 90

157. The cheapest fertilizer is the one in which one dollar purchases the greater amount of plant-food. In No. 1, $29 obtained $13.80 worth, which is at the rate of 48 cents worth for $1. In No. 2, $38 buys $24.90 worth of plant-food, or at the rate of 65 cents worth for the dollar. The difference between the commercial value, as calculated, and the selling price, is to cover expenses of manufacture, bagging, shipping, commission fees, and profits.

6b. Advice as to their use

158. We have seen that plants must have all three of the general fertility elements—nitrogen, phosphoric acid, potash—in order to thrive. It frequently occurs, however, that the soil is rich enough in one or two of them; and in that case, it is not necessary to apply all of them.

159. If a liberal application is made of one element, the plant must use more of the other elements which are already in the soil, in order to balance up its growth. It may result, therefore, that the addition of one element exhausts the soil of some other element. For example, if heavy growth is obtained by the addition of nitrogen, the plant may need to draw so heavily upon the stores of available phosphoric acid as to deplete the soil of that material.

160. Again, no results can be obtained from the addition of one element unless the other two are present in sufficient quantity. In general, therefore, it is safer to apply complete fertilizers.

161. Yet, in some cases, it is unwise to apply complete fertilizers. This is particularly true of the application of nitrogen. The growth may already be so heavy that the addition of nitrogen would cause an overgrowth, and yet the plants may need fertilizing. This danger of too much growth is greatest with fruit plants (114).

162. If nitrogen conduces especially to leaf growth (134), then it must be the element which is most important in the fertilizing of the vegetables which are grown for their leaves or succulent stalks, as rhubarb, cabbage, lettuce, spinach, asparagus; and it is also very important in the growing of hay and succulent fodder.

163. Nitrogen leaches rapidly, especially if applied in the form of nitrate of soda or sulfate of ammonia. It is, therefore, advisable to apply it in the spring; and when used in liberal amounts, it should be applied at intervals, and not all at one time.

164. Phosphoric acid and potash, even if soluble, do not leach badly, as a rule, because

they tend to form insoluble compounds with soil constituents. The more vegetable matter a soil contains, the less pronounced is the action of leaching. As a rule, commercial fertilizers are applied after the ground is fitted, and then harrowed in or drilled in.

165. The amounts and kinds to apply are determined by (*a*) the analysis of the material (that is, its richness in plant-food), (*b*) its cost, (*c*) the richness of the soil in plant-food, (*d*) the tilth or texture of the soil (60, 49*a*), (*e*) the kind of crop, (*f*) the kind of farming, whether intensive or extensive (111*a*, 111*b*). It follows, therefore, that the mere analysis of the soil and the plant cannot determine what fertilizer it is most profitable to use.

166. What fertilizers to use, and how to apply them, are subjects which are discussed in bulletins and books by many authors; but even after reading all the literature, the farmer must experiment with his own land and his own crops, to determine just what materials are most profitable for his use. In other words, the advice as to fertilizers is more valuable in teaching a man principles, in suggesting means of experimenting, and in designating the probabilities of any line of action, than in specifying just what fertilizers one shall use. An area on one side of a field may be devoted to such

experiment, on different parts of which the various elements and combinations of them may be applied.

SUGGESTIONS ON CHAPTER VI

127*a*. An element is a simple substance. It is not made by a combination of any other substances, and by no known means can it be separated into any other substances. Sulfur, nitrogen, and phosphorus are elements. The known elements number about 70.

127*b*. The elements are represented by one or more letters, called symbols. Usually the first letter of the name is employed. Thus, nitrogen is designated by N, phosphorus by P, sulfur by S. When the names of different elements begin with the same letter, as sulfur and sodium, this rule cannot be followed. In such cases, letters from the name of one of the elements in some other language are used. Thus, Na is used for sodium, natrium being the Latin of sodium. Similarly, P might represent phosphorus or potassium; hence K is used for potassium, which in Latin is kalium.

130*a*. Compounds result from the chemical union (30*c*) of two or more elements. The compound may not resemble in any way any of the elements contained in it. The proportions in which elements unite vary, and the same elements may be made to unite in different proportions. The same compound always contains the elements in exactly the same proportion.

130*b*. Compounds are represented by writing together the symbols of the elements composing them, together with figures to represent the proportions. Thus, potash, K_2O, is a compound of two parts of potassium and one of oxygen, O. Lime, CaO, is composed of the elements calcium, Ca, and oxygen, and its chemical name is calcium oxid. Other compounds are nitrate of soda, $NaNO_3$; ammonia, NH_3 (H representing the element hydrogen); water, H_2O; sulfuric acid, H_2SO_4; ammonium nitrate, NH_4NO_3; ammonium sulfate $(NH_4)_2SO_4$ (the NH_4 being taken twice); starch, $C_6H_{10}O_5$ (C; representing carbon); salt, NaCl (Cl standing for chlorin).

130*c*. Phosphoric acid and potash are not elements, but compounds. The elemental forms are phosphorus and potassium. It is customary, however, to speak of nitrogen, phosphoric acid and potash as the elements of plant-food. Here the word element is not used in the chemical sense, but rather as the simplest form in which plants can use these substances.

131*a*. Roots have the power of dissolving plant-food (30, 30*a*), but this is only a process of making it soluble. Substances which are not soluble in rain water may be soluble in soil water, for the water in the soil contains various acids. Even when a substance is in solution, the plant has the power of rejecting it; it is thereby not available as plant-food. For example, nitrogen in the form of nitrites (as nitrite of soda, $NaNO_2$) is not available, although it is soluble; but nitrogen in the form of nitrates (as nitrate of soda, $NaNO_3$) is available. Charcoal is not available plant-food, although it is carbon, and carbon enters more largely than any other element into plant tissue. But when the charcoal is burned, it forms a gas called carbon dioxid or carbonic acid (CO_2), from which the plant can get carbon.

140*a*. The black or blue head of an old-fashioned sulfur match is a paste containing the element phosphorus, P. On igniting the match, this phosphorus unites with the element oxygen, O, in the air to form a small white cloud, which is the compound phosphorus pentoxid. Its symbol is P_2O_5, which means that it is made by the union of two parts of phosphorus and five parts of oxygen. Phosphorus pentoxid is known in agriculture as phosphoric acid.

143*a*. The term superphosphate is sometimes used in the same sense as acid phosphate; that is, to designate available phosphates, or those which are made up of monocalcic and dicalcic phosphates. A fertilizer containing available phosphoric acid, but no nitrogen or potash, is often called a plain superphosphate. Complete fertilizers contain all three of the important plant-foods.

153*a*. Moisten a strip of blue litmus paper with vinegar or sour milk, and note the change in color. Then add to the milk or vinegar some lime water till it no longer tastes sour, and again try the litmus paper. It will no longer turn red. Try some air-slaked lime in the same way. Make the same test with plaster of paris or gypsum, which is sulfate of lime. This will not neutralize the acid or sweeten the milk or vinegar. Make the same test with salt and sugar. A substance, which turns blue litmus red is acid; one which turns red litmus blue is alkaline.

166*a*. The experiment stations of most of the older states issue bulletins of advice on the use of fertilizers, and these should be studied. In many states there are laws designed to protect the purchaser of fertilizers; and fertilizer control stations are established to analyze the different brands and to publish the results. The general subject of fertilizers is presented in Voorhees' book on "Fertilizers." Good advice will also be found in Chapter xii. of Roberts' "Fertility."

166*b*. Every school should have bottles of the leading fertilizer chemicals for exhibition; as muriate and sulfate of potash, kainit, gypsum or plaster, bone and rock phosphates, bone-black, dried blood, nitrate of soda, sulfate of ammonia, air-slaked lime, and quick-lime. These can be obtained from dealers in fertilizers.

Part II

THE PLANT, AND CROPS

Chapter VII

THE OFFICES OF THE PLANT

1. *The Plant and the Crop*

167. In an agricultural sense, the plant, as a representative of the vegetable kingdom, has four general types of uses, or fulfils four offices: it aids in the formation, maintenance and improvement of soils; it influences the climate and habitableness of the earth; it is the ultimate source of food of domestic animals; it, or its products, may be of intrinsic value to man.

168. When plants are grown in quantity, they, or their products, constitute a crop. This crop may be the produce of a bench of carnations, a field of barley, an orchard of peaches, a plantation of tomatoes, or a forest. The crop may be grown for its own or intrinsic value, or for its use in preparing the land for other crops.

2. *The Plant in its Relation to Soil*

169. The plant is a soil maker. It breaks down the rock by mechanical force and by dissolving some of its constituents (30, 30*b*). It fills bogs and lagoons and extends the margins of lakes and seas (32, 32*a*).

170. The plant is a soil improver. It opens and loosens hard soils, especially if, like the clover, it has a tap-root, which it sends deep into the earth. It fills and binds loose and leachy soils. When it decays it adds humus (33, 34, 73, 74).

171. The plant is a soil protector. It prevents the washing of soils, and protects the sands of dunes and shores from the winds. It holds the rainfall until it soaks into the soil (70, 116).

3. *The Plant in its Relation to Climate*

172. The plant influences the moisture supply: by modifying the distribution of precipitation; by causing the retention of the precipitation; by lessening evaporation; by adding moisture to the atmosphere.

173. The plant influences the habitableness of the earth by other means: as by modifying extremes of temperature; by affording wind-

breaks; by supplying shade; by contributing to the beauty and variety of the landscape.

4. *The Plant in its Relation to Animal Life*

174. Nearly all domestic animals live directly on plants. These are herbivorous animals, such as cattle, horses, sheep. But even the flesh which carnivorous animals eat—as dogs, cats—is directly or indirectly derived from herbivorous animals; for "all flesh is grass."

175. The round of life begins and ends with the soil. The soil contributes to feeding the plant, the plant feeds the animal, and the animal passes at last into the soil. In this round, there is no creation of elements, and no loss; but there are endless combinations, and these combinations break up and pass away. To raise the plant, therefore, is the primary effort in agriculture.

5. *The Plant has Intrinsic Value to Man*

5a. *As articles of food or beverage*

176. Plants or plant-products may be staples or necessaries, as wheat, rice, potatoes, beans; semi-staples, or articles of very general and common use, as apples, oranges, buckwheat;

luxuries or accessories, as quinces, cauliflowers, glass-house vegetables; condiments, as spices; beverage products, as cider, wine.

177. Plants or plant-products may be food for animals, as grains, ground feed, fodders, forage or field pasturage.

5b. As articles used in the arts

178. Plants may afford textiles or fibers, as cotton, hemp, flax, jute; wood, lumber and timber; medicines, as quinine, opium, ginger.

5c. As articles or objects to gratify æsthetic tastes

179. Plants are the source of most perfumery, and of many dyes and paints.

180. Plants are themselves useful as ornamental subjects. They may be grown for their effects as individuals or single specimens, as a tree, a shrub, or a plant in a pot; or for their effects in masses in the landscape.

181. Plants are useful for their flowers or ornamental fruits. The flowers may be desired in mass effects, as single specimen plants, or as cut-flowers. The growing of plants for their effects as individuals or for cut-flowers is floriculture; the growing of them for their combined or mass effects in the open (or on the lawn) is landscape horticulture (9).

SUGGESTIONS ON CHAPTER VII

170*a*. Tap-roots (Fig. 33) extend the benefits of root action to great depths. They drain, aerate and comminute the soil;

Fig. 33. The deep root-system of red clover.

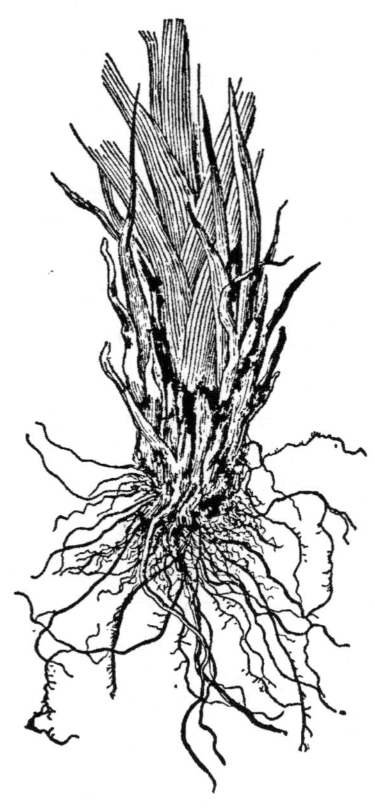

Fig. 34. The shallow root-system of orchard grass.

and the plant-food which they bring from the subsoil is left, when the plant decays, in such place and condition that surface-rooted plants can get it. With the clover, compare a grass (Fig. 34).

171*a*. In many countries definite efforts are made to hold

loose sands from drifting by winds, as along the coasts of the sea. Sand-loving plants with strong running roots or rootstocks—as various grasses and sedges—are used for this purpose. One of the uses of windbreaks is to lessen the drifting of sands. Bluffs and railway embankments are often held from caving and washing by means of strong-rooted plants.

172*a*. Large forests probably have some influence in distributing the rainfall, the precipitation tending to be greatest near the forest areas. By some persons it is thought that the total precipitation is increased by forests, but this point is in dispute. The off-flow or outflow from forest-covered, or from any plant-covered, lands is more gradual than from bare lands; thus floods are more frequent and more serious the more completely the forests are removed. This is illustrated in the floods on the Ohio and other rivers.

172*b*. Plants lessen evaporation chiefly in the capacity of shelter-belts. Windbreaks check evaporation from adjacent lands (see King, "The Soil," pp. 204–206); and this is one valuable effect of windbreaks for fruit-plantations in dry climates (see Bailey, "Principles of Fruit-Growing," pp. 48–51.) Forest areas contribute some of their moisture to the atmosphere of contiguous areas; and plants give off moisture from their growing parts.

173*a*. For a full discussion of windbreaks, see "Principles of Fruit-Growing," pp. 47–57, 62–92.

CHAPTER VIII

HOW THE PLANT LIVES

B. M. DUGGAR

1. *The Plant Activities*

182. The plant is a very dependent structure: it must be supplied with water and certain soluble salts from the soil, oxygen and carbon dioxid from the air, in addition to sunlight and a certain amount of heat. When these conditions are fulfilled,—somewhat as a plant's ancestors have been accustomed to them,—the plant must grow, provided no extraneous diseases or accidents overtake it.

183. A growing plant is influenced by all of the external conditions about it,—it is sensitive, or irritable. In studying growth processes, we must remember that these processes are occurring in a highly irritable living object, and that they cannot be explained by physics and chemistry alone. To study how a plant lives, then, one must consider the factors of growth, actual growth processes, and the conditions to which growth is sensitive.

2. *The Factors of Growth*

2*a*. *Water in the plant*

184. The rigidity or stiffness of any herb or succulent part is largely dependent upon its water content. If a succulent branch is severed, it soon loses its water by evaporation, and it becomes flaccid, or wilted. The proper extension, or turgidity, of the cells of plants with water is necessary for active growth. The passage of the soil water into the plant, and thereafter its transfer from cell to cell, is accomplished by the principle of osmosis, which is the diffusion of liquids through membranes.

185. Surrounding each rootlet for some distance back of the tip is a fringe-like growth of delicate root-hairs. These hairs are single, tubular cells, the outgrowth of single cells in the outer layer of the root. Each one contains within its walls, as do all active cells, living matter called protoplasm, along with cell-sap. In the soil these delicate hairs pass readily amongst the soil particles, covering an immense amount of space. Owing to the dense cell-sap of the root-hairs these hairs absorb water by osmosis. In solution in the soil water are minute quantities of food substances, and each of these is absorbed independently according

to certain physical laws. The absorptive activity of the root-hairs is so great that water may be extracted from a soil apparently dry.

186. Plants contain much water; but it requires oven temperatures to drive out all the water from plant substance. The total water in some plants, as determined by the chemist, is as follows:

Dry clover seed	6.4	per cent.
Dry beans	12.5	" "
Green apple twigs	50.0	" "
Potato tubers	80.0	" "
Green clover tops	85.0	" "

187. In order to secure the proper amount of food substances, water is absorbed in greater quantity than can actually enter into the composition of the living plant; and the surplus water is thrown off by a process of evaporation known as transpiration. The water is rapidly transpired from living plant surfaces, especially from the leaves and green stems.

188. To aid transpiration, the leaves are provided with thousands of minute pores in the epidermis, connecting with the delicate tissues within. These pores, or stomata, are especially abundant on the under surfaces of leaves. With changes in the humidity, these stomata open or close, to facilitate or inhibit transpiration. Like evaporation, transpiration is has-

tened by higher temperatures, dry air, wind, and the movements of the plant. On a very hot day, or with insufficient soil moisture, a plant may wilt, due to the fact that all of the facilities for checking transpiration fail to keep the balance between root absorption and transpiration. The plant gives off more water than it takes up; therefore, it wilts.

189. The absorptive activity of the roots gives rise to a pressure which tends to force the current upward. In fact, root-pressure, together with other forces, especially transpiration, causes the crude sap to ascend through the woody bundles of the plant; and by means of these bundles absorbed solutions are carried upward through all parts of root and stem, and through the leaf-stalk, veins and veinlets to all parts of the active leaf surface.

2*b*. *Soluble salts from the soil*

190. Along with the soil water absorbed by the roots, minute quantities of the various mineral salts necessary for plant growth are taken in. These salts are in solution. In the plant, these solutions become a part of the ascending sap, and they are diffused to all parts where assimilation goes on. Many soil elements not utilized by the plant are also absorbed in small quantities,

but not being used, a certain equilibrium is established, and no more is absorbed. Carbonic acid and perhaps other substances excreted by the root aid in dissolving some of the mineral salts (30).

191. Various substances are taken in with the soil water. Sodium and potassium nitrate (nitre), calcium phosphate (phosphate of lime), and potassium sulfate are well known ingredients of fertilizers. Chemical analysis and experiments show that from these and allied salts the plant obtains from the soil such necessary elements as nitrogen, potassium, phosphorus, calcium, and sulfur. In addition, plants also secure from the soil traces of iron, and whatever magnesium, silicon, and other elements may be necessary.

192. When a plant is burned in air, the ash contains all of the above-named elements except the nitrogen and a part of the sulfur and phosphorus. Nitrogen, one of the most important of plant-foods, can be used chiefly in the form of nitrates, except in the case of leguminous plants (110, 138), in which it is also taken from the air in some obscure way by the root tubercles.

2*c*. *Oxygen*.

193. Oxygen is essential to all of the life processes in the plant, as well as to the animal. For perfect germination oxygen is required, and this gas is absorbed by all living or growing plant

surfaces, but more abundantly by the delicate parts and by the stomata of the leaves. Entering these stomata, it is readily diffused throughout the structure of the plant.

194. Oxygen is constantly absorbed, and associated with this absorption is the giving off of carbon dioxid. This appropriation of oxygen and escape of carbon dioxid are results of respiration, a process equivalent in its purpose and results to respiration in animals. Young growing plants absorb an amount of oxygen about equal to their volume, in from twenty-four to thirty-six hours. Germinating seeds absorb oxygen, and give off about an equal quantity of carbon dioxid.

195. Plants, or parts of plants, which have been injured, and those parts in which decay is imminent, as in fading flowers, respire more rapidly than normal parts. Respiration practically represents the constant breaking down of living substance used in the vital processes.

196. Oxygen is also taken in through the roots. Land plants, whose roots are deprived of their air by too much water, are soon suffocated. This is especially noticeable in a field of Indian corn or maize which has been overflowed; and it is also a condition frequently met with in those greenhouses where an abundant use of water is the first rule. Many plants which have

become accustomed to boggy regions, and many greenhouse plants, send up to the surface numerous root formations for the special purpose of securing fresh air, or oxygen.

2d. Carbon dioxid and sunlight

197. The element which is present in greatest amounts in plants is carbon. This material is derived from the carbon dioxid (or carbonic acid gas) of the air.

198. In order to become plant-food, the carbon dioxid of the air is first absorbed; largely by the leaves; and then its utilization depends upon the green coloring matter of leaves,—or the chlorophyll,—and upon sunlight. The chlorophyll absorbs some of the energy of sunlight, and by means of the energy thus provided, the carbon unites in some obscure way with the elements of water, ultimately forming starch, and oxygen is thrown off. This process of the formation of plant-food from carbon dioxid and water, with the consequent giving off of oxygen, is photosynthesis (sometimes known as carbon assimilation). It is the reverse of respiration, in which oxygen is taken in and carbon dioxid given off.

199. During the day much more oxygen is given off as a result of photosynthesis than there is oxygen used in respiration, so that oxy-

gen is given off in great excess of carbon dioxid, and plants are said to purify the air. At night, no carbon assimilation goes on, and the effect of respiration is to give off small quantities of carbon dioxid.

2e. *Heat, or a definite temperature*

200. Heat increases the absorptive activity of the roots, the rate of transpiration, the amount of respiration, and assimilation of carbon.

201. A more or less definite degree of heat is necessary for all living processes. As a rule, seeds will not germinate at the freezing point, and all growth is suspended at that temperature. Plants grow best within a very small range of temperature, known as the optimum temperature. The greatest amount of leaf surface compatible with proper seed production will be found at this temperature. As a rule, other conditions being equal, plants of tropical regions are succulent, and green tissues preponderate. In temperate regions there is the highest development of woody structures combined with leaf development. In the frigid regions the softer green parts are greatly reduced, and, while the woody portion is of less extent than in the temperate regions, relatively it preponderates.

202. Different plants are injured by different

temperatures. Plants like the cotton and the melon are killed by a temperature several degrees above freezing. The living protoplasm is stimulated to give up its water, the roots are chilled and cannot supply to the leaves that water necessary to offset transpiration, and, as a result, the leaves soon wilt and blacken. On the other hand, even the green parts of some plants will withstand freezing temperatures. The ability to resist cold depends upon the constitution of the cell-sap and the power of the protoplasm gradually to free itself of surplus water.

3. *The Processes of Growth*

203. The starch which results from the assimilation of carbon is stored in the leaves during the day, and at night it may be entirely removed after being converted into a soluble substance, sugar. Some of this soluble substance unites with elements taken from the soil to form more complex compounds used in growth, and some of it is again converted into starch and stored in tubers, stems, or thickened leaves, for future growth purposes.

204. The external evidences of growth are changes in form and size of the different parts. The internal evidences of growth are to be seen in the individual cells of which the plant is com-

posed,—new cells are made, and others are modified in size or form. It is probably impossible for a plant to live without growing; but the growth may be so slight that the plant is no longer of any use to the farmer.

205. The young stems of plants elongate throughout the entire length of the growing portion. But the lower part soon reaches the limit of its growth, the rear internode—or space between the joints—ceases to elongate, and further growth proceeds in the newer parts above. That is, while there is an elongation or stretching of the shoot itself, this elongation gradually lessens below, so that the region of most rapid growth is constantly in the freshest and softest part of the shoot. Notice that the distance between the joints in growing shoots tends to increase.

206. The root grows differently. The tip of the growing root is hard, being protected by what is known as a root-cap. Growth in length takes place just behind this hard tip, not throughout the length of the growing part. The root, therefore, is able to pick its way around obstacles, since its growth takes place practically at its end.

207. In most of our woody plants, increase in diameter is effected by a layer of growing tissue, the cambium, located just beneath the bark; and every year it gives rise to a new layer of wood on

the outside of the old wood, and to a new layer of bark on the inside of the old bark. Thus the heart-wood is the oldest wood, and the outside bark constantly breaking off is the oldest bark. The interior wood takes less and less part in the activities of the plant, and the heart-wood of trees is nearly useless except as a support to the plant.

4. *Irritability*

208. Growing parts are sensitive or irritable. This irritability is shown in definite movements, or growth reactions, under the influence of the forces which irritate them.

209. Some plants make visible movements, and may even be sensitive to shocks. The sensitive-plant suddenly closes its leaves and droops when touched; the leaves of sun-dew and other insectivorous plants close tightly upon their prey; and the searching tendril of the cucumber or gourd gradually bends around the object it touches.

210. Green parts turn towards the light, and assimilation is thereby increased. Plants in windows turn the broad surfaces of their leaves perpendicular to the incoming rays of light; and a seedling grown under a box into which light is admitted through a single slit will grow

directly towards that slit, and even through it to the brighter light.

211. Plants are sensitive to gravitation. The first root of the germinating seed is so sensitive to gravity that it ordinarily grows downward, wherever it may be and whatever may be its position. On the other hand, the first shoot is oppositely affected by gravity, and if a potted seedling is placed horizontally the stem soon directs itself upward. While its general tendency is downward, the root is nevertheless attracted in any direction by the position of the greatest food-supply.

212. The reactions of plants to their environments or surroundings may cause the plants to vary, or to assume new forms or characteristics; and these new features may be of use to the farmer. Thus, with more light, the better are the roses or carnations grown under glass; the richer the soil, the stronger is the growth; the higher the altitude or latitude, the greater is the proportion of dwarf plants.

SUGGESTIONS ON CHAPTER VIII

182*a*. A salt is the substance formed from the union of an acid with some inorganic substance or base. The salt may be neutral, — neither acid nor alkaline. Thus sulfuric acid and lime form the salt sulfate of lime or gypsum; nitric acid and soda form the salt nitrate of soda; muriatic (hydrochloric) acid and potash form muriate of potash; muriatic acid and soda form muriate of soda, which is commonly known as salt,—that is, it is common salt.

184*a*. From a potato tuber which has lain in the air until somewhat wilted, cut circular segments about one-fourth of an inch or less in thickness. Place some of these pieces in water, and others in strong salt solution. In a short time those in water become more rigid, while those in strong salt water become flaccid. The cell-sap of the potato, containing some salts and sugars in solution, is a denser solution than the water, and the flow of water is inward to the denser solution; hence the pieces absorb water. Of those pieces in strong salt solution the flow of water is outward, and the potato segments lose some of their water and become flaccid. See Atkinson's "Elementary Botany," pp. 13–18.

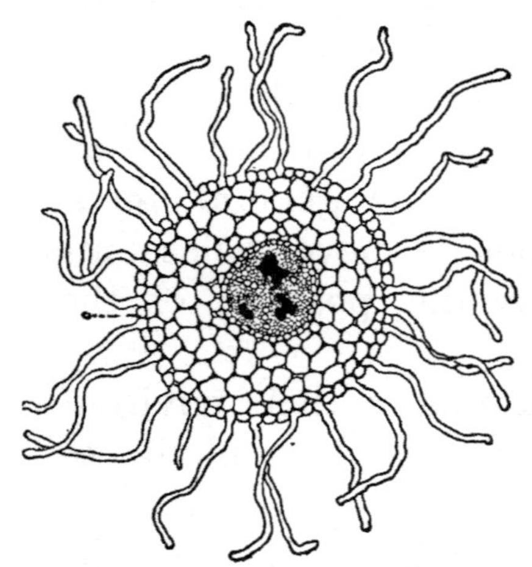

Fig. 35. Root-hairs, enlarged.

185*a*. A cross-section of a rootlet in Fig. 35 shows the root hairs. These hairs are seen to be prolongations of the outer or epidermal cells.

185*b*. By germinating a bean, pumpkin seed, or wheat in moss, or between folds of moist thick cloth, the root-hairs may be observed. Fig. 36 shows the fringe of hairs on such a seedling; and Fig. 37 shows how the root-hairs attach the soil particles to the root. For a longer account of root-structures and root-action, compare Sorauer, "Physiology of Plants for the Use of Gardeners," pp. 4–7.

186*a*. Any one who has handled both green and dry fodder has a general idea of how much water there may be in plants. Why do apples and grapes and cabbages shrivel after they are picked?

188*a*. A single breathing-pore is a stoma or stomate. The

plural is stomata or stomates. Fig. 38 shows a fragment of leaf in cross-section, *a* being a stoma opening out on the lower surface. Looking down upon the peeled-off epidermis of the lower surface, stomata are seen at Fig. 39.

188*b*. Cut off a leafy branch of any herb, insert the stem through a perforated cork into a bottle of water, and then place the whole under a bell-glass. Note how soon the water vapor thrown off condenses upon the glass. Compare Fig. 10, page 58.

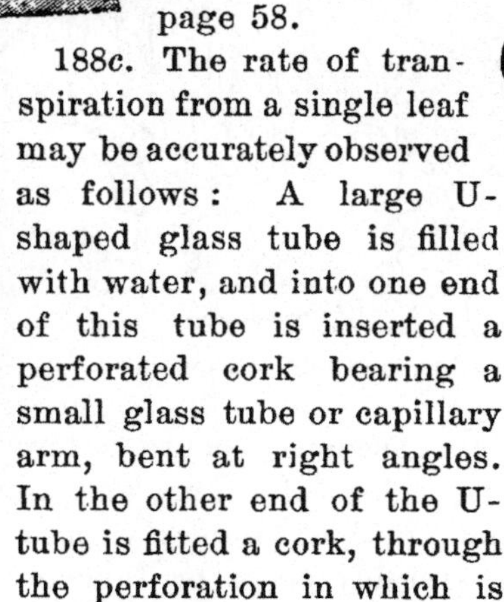

Fig. 36. The root-hairs as seen on a dark, damp cloth.

188*c*. The rate of transpiration from a single leaf may be accurately observed as follows: A large U-shaped glass tube is filled with water, and into one end of this tube is inserted a perforated cork bearing a small glass tube or capillary arm, bent at right angles. In the other end of the U-tube is fitted a cork, through the perforation in which is inserted the leaf-stalk, with the stem reaching the water, as shown in Fig. 40. When this last cork is forced in, water will fill the capillary arm; and the recession of the water in this arm to supply that transpired shows the rate of transpiration. Wax or paraffin should be used to seal around the perforations.

189*a*. Root-pressure may be made evident by a very simple experiment. An inch or so above ground, cut off a stem of some actively-growing herbacous plant like the sunflower. Fit tightly

Fig. 37. How the soil adheres to the young root.

over this stub a few inches of rubber tubing, partially filling the tubing with water, and into the free end fit closely a small glass tube several feet long, fastening the tube perpendicularly to a stake. In a few hours water will begin to rise in the glass tube. Root-pressure in the common nettle will sustain a column of water over ten feet in height, and in the grape vine a column more than thirty feet in height.

189*b*. The sap ascends through the young woody parts,—the sap-wood in our common trees,—not between the bark and wood, as commonly supposed. To note the special channels

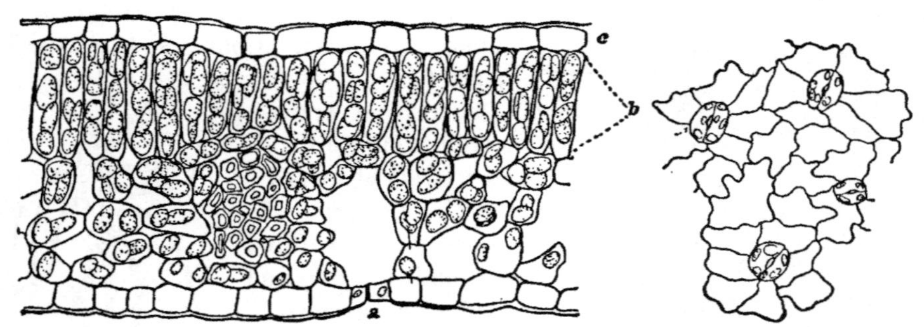

Fig. 38. Cross-section of a leaf. Stoma at a. Fig. 39. Four stomata.

through which sap ascends, secure a few joints of green corn, a blade of celery, a leaf of canna, and some woody branch, and put the stem ends into a tumbler with a solution of some red dye or stain, preferably eosin or fuchsin. Often in the course of a few hours there is external evidence that the colored liquid ascends through definite channels, at least with the succulent herbs. Now cut off the stems and note the colored regions,—in the corn those thread-like groups of fibers so noticeable when an old cornstalk is broken ; in the celery, likewise, through those stringy fibers known to all who have eaten tough celery ; and in woody plants, through the layers of wood nearest the bark.

190*a*. For fuller discussions of the subjects outlined in 190 and 191, consult Sorauer, "Physiology of Plants for the Use of Gardeners," pp. 30–44, 48–51.

194*a*. To show that carbon dioxid (or carbonic acid gas) is given off, fill a large-mouthed bottle half full with beans or peas which have been soaked for a day, add a small quantity of water, and cork it. After twenty-four hours, pass a lighted wax taper or waxed cord into the jar, and it will be extinguished. A small vessel of lime water inserted into the bottle will show some turbidity, or a slight pre- for carbonic acid gas. These tests for carbonic acid. These tests indicate that carbon dioxid has replaced the oxygen of the jar. Make the same tests in a jar of air, and see that the taper burns and that the lime water is not made turbid.

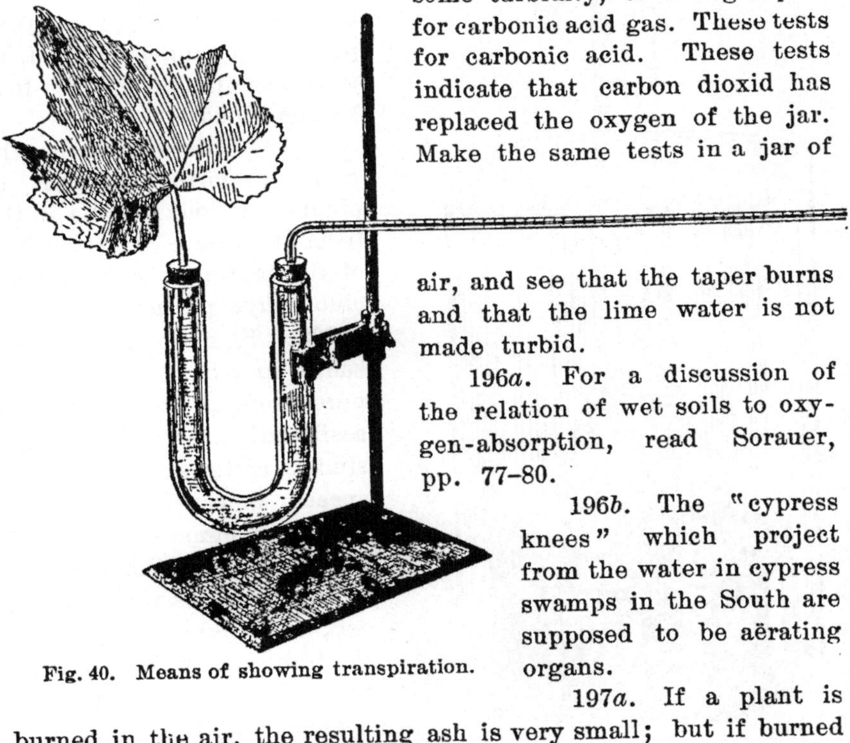

Fig. 40. Means of showing transpiration.

196*a*. For a discussion of the relation of wet soils to oxygen-absorption, read Sorauer, pp. 77–80.

196*b*. The "cypress knees" which project from the water in cypress swamps in the South are supposed to be aërating organs.

197*a*. If a plant is burned in the air, the resulting ash is very small; but if burned without free access of air, as in a charcoal pit, there remains a charred mass almost as great in volume as the substance burned. This mass is largely carbon, a most important element in all living matter, or protoplasm. In combination with the elements of water, carbon also forms most of the cellular tissue of plants, likewise the starches and the sugars, all of which are called carbohydrates. The manufacture of these starch-like compounds by the appropriation of the carbon dioxid of the air is one of the peculiarities of plant-life; and animals depend upon plants for the preliminary preparation of these necessary compounds.

198*a*. The word assimilation is used in a very restricted sense in plants, as defined in 198. In general speech it means the appropriation of prepared or digested food, as the assimilation of the food by the blood

Fig. 41. Experiment to show the giving off of oxygen.

198*b*. Chlorophyll is the green coloring matter of plants. It looks to be in the form of minute grains. Most of the cells in Fig. 38 contain chlorophyll grains.

198*c*. Plant-food, in the sense in which the term is here used, is the product of assimilation,—starch or some similar material. In common speech, however, the term is used to designate any material taken in and ultimately used by the plant, as nitrates, potash, water; and this use of the term is so well established that it cannot be overthrown.

198*d*. For further light on assimilation, compare Arthur and MacDougal, "Living Plants and their Properties," pp. 145–152, and Atkinson's "Elementary Botany," Chapter x.

199*a*. Place under a funnel in a deep beaker, containing spring or stream water, fresh bits of water-weed (*Elodea*, or *Anacharis*, *Canadensis*), and invert over the end of the funnel a test-tube filled with water, as in Fig. 41. In the sunlight bubbles of gas will be seen to rise and collect in the tube. If a sufficient quantity of this gas could be collected, on testing it with a lighted taper the flame would be seen to quicken per-

Fig. 42. Opening of a bud of pear.

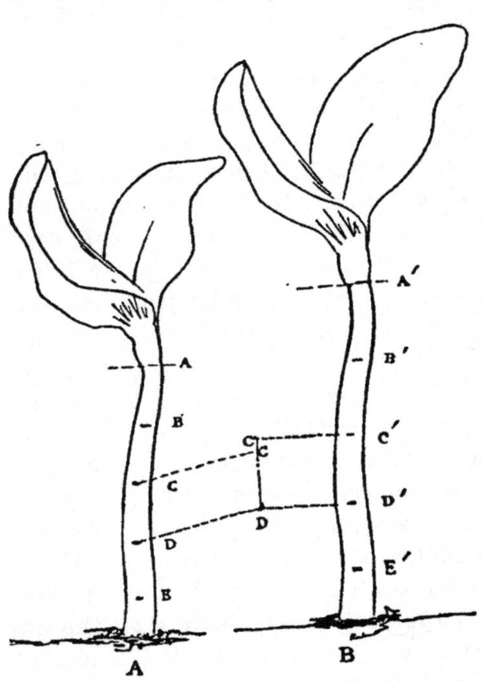

Fig. 43. The marking of the stem and the spreading apart of the marks.

ceptibly, indicating more oxygen than is contained in the air. In this case the carbon dioxid used is in solution in the water. The Elodea is common in still ponds.

201*a*. On the subject of temperature and plant life, compare Bailey, "The Survival of the Unlike," pp. 44–48, Chapters xvii. and xix.; and Chapter xiii. of Gaye's "Great World's Farm."

202*a*. Compare Arthur and MacDougal, "Living Plants and their Properties," pp. 85–98, for a discussion of the influence of cold in injuring plants.

203*a*. To test for starch in a potato tuber or other storage organ, spread a drop of tincture of iodine on the cut surface, and the blue or violet color indicates the presence of starch. Test the laundry starch.

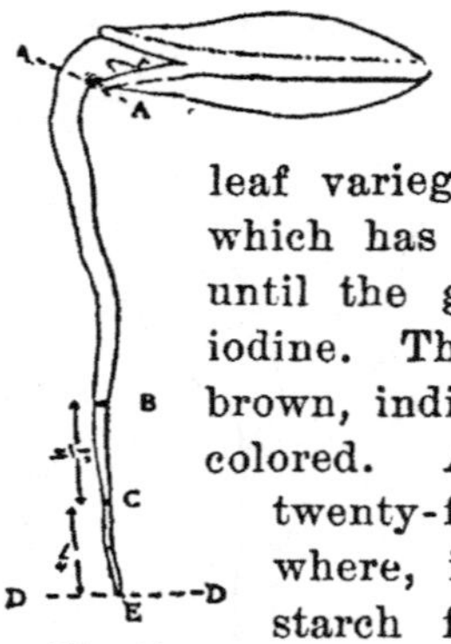

203*b*. To determine that starch is formed only in the green parts of leaves, secure a leaf variegated with white, like a coleus or geranium, which has been in sunlight. Place it in hot alcohol until the green color disappears, and then add some iodine. The parts which were green are colored violet-brown, indicating starch, but the white parts are uncolored. Another leaf covered with dark cloth for twenty-four hours will show little or no starch anywhere, indicating the removal in darkness of the starch formed in sunlight.

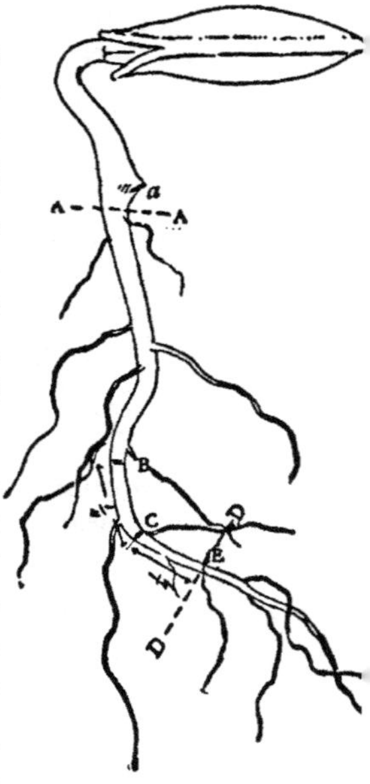

Fig. 44. Marking the root.

204*a*. The opening bud of a beech is a good example for observation of growth, as it expands from day to day. The long scales of the winter bud become looser, and gradually, by the elongation of parts between them, the scales are forced apart, showing at the base of each a minute leaf of perfect form. Daily the leaf increases in size, the internodes or stem portions between the leaves elongate, the scales fall away, and from a bud of an inch in length, by elongation throughout its whole extent we have a leafy twig of many inches, with a terminal bud, and a bud in the axil of each leaf. The beginning of the spring growth is likewise well shown in the pear bud, Fig. 42. Consult Bailey's "Lessons with Plants," pp. 44–72, for fuller discussions, with many illustrations, of the opening of buds.

Fig. 45. The root grows in end portion.

205*a*. Mark a young stem, as at A in Fig. 43; but the next day we shall find that these marks are farther apart than when we made them (B, Fig. 43). The marks have all raised themselves above the ground as the plant has grown.

The stem, therefore, has grown throughout its length rather than from the end.—*Bailey, "Lessons with Plants," p. 322.*

206*a*. Pull up a squash-seed when it has sent a single root about two inches deep into the earth; or better, germinate it between layers of blotting-paper or cloth. Wash it very carefully, if dirty, and lay it upon a piece of paper. Then lay a rule alongside of it, and make a mark (with indelible ink) one-quarter of an inch, or less, from the tip, and two or three other marks at equal distances above (Fig. 44). Now carefully replant the seed. Two days later, dig it up; we shall most likely find a condition something like that in Fig. 45. It will be seen that the marks E, C, B, are practically the same distance apart as before, and they are also the same distance from the peg, A A. The point of the root is no longer at D D, however, but has moved on to F.—*Bailey, "Lessons with Plants," p. 321.*

207*a*. We now see that the sap of trees is a very complex substance. It is the juice or liquid in the plant. When this liquid first comes in at the root it is water, containing very dilute solutious of various substances. But the sap also carries the products of assimilation to all parts of the plant, to build up the tissues. In common speech, the upward-moving water, recently taken in from the soil, is often called crude sap; and the liquid carrying sugars and other organic compounds is called elaborated sap. Botanists do not often use the word sap, but speak of the water or fluids of the plant.

209*a*. See the discussions and pictures of moving parts in Bailey's "Lessons with Plants," pp. 396–406; also Barnes' "Plant Life," pp. 188–208; Atkinson's "Elementary Botany," pp. 82–92; Arthur and MacDougal's "Living Plants," Chapters i.–iv., and other botanical treatises.

CHAPTER IX

THE PROPAGATION OF PLANTS

1. *The Kinds of Propagation*

213. Plants naturally propagate by two general means,— by seeds and by buds. All the modes of the propagating of plants employed by the farmer and gardener are but modifications of these two general types.

214. The farmer has three objects in view in the propagation of plants: to renew the generation, or to prevent the stock from dying out; to increase the number of plants; to perpetuate a particular variety. Thus, the farmer must resow his wheat, or he will lose the stock; but he expects to secure more plants than were concerned in the production of the seed which he sows; and he also expects to reap a particular variety, as Diehl or Mediterranean.

215. Seeds are always able to preserve the race or stock and to increase the number of plants, but they are not always able to produce the variety which bore them. Most farm crops and most garden vegetables reproduce the va-

riety from seeds; but most fruits and trees and shrubs do not, and in such cases recourse is had to bud propagation, as layers, cuttings, grafts.

2. *Seedage, or Propagation by Seeds*

2a. *Requisites of germination*

216. In order that seeds shall germinate, the seeds themselves must be viable (or "good"). Viability depends upon (*a*) the maturity of the seeds, (*b*) freshness,—they shall not have lost their vitality through age,—(*c*) the vigor and general healthfulness of the plant which bore the seeds, (*d*) proper conditions of storage.

217. (*b*) The length of time during which seeds retain their vitality varies with the kind of plant and with the conditions under which the seeds were grown. That is, there is a normal vitality and an incidental vitality. Most seeds germinate best when not more than one or two years old, but retain strong vitality three or four years; but some seeds, notably those of onions and parsnips, are usually not safe after a year old.

218. In order that seeds shall germinate, they must also have proper surrounding conditions: moisture, free oxygen (air), warmth.

219. The ideal condition of the seed-bed, so far as water is concerned, is that it shall be moist, not wet. Wet soil injures seeds, largely by excluding oxygen. The older and weaker the seeds, the greater is the necessity for care in applying water: they should be kept only slightly moist until germination is well started. The soaking of seeds starts the germinating processes, but it should not be continued above twenty-four hours, as a rule, and should not be employed with very weak seeds.

220. Oxygen is supplied to germinating seeds if sufficient air is allowed to reach them; and the air reaches them if they are not planted too deep, nor kept too wet, nor the soil allowed to "bake." But all these conditions are greatly modified by the kind of soil.

221. For each kind of seed there is a certain degree of warmth under which it will germinate to the best advantage; and this is called the optimum temperature for that seed. The optimum temperature is not uniform or exact, but ranges through a limit of five to ten degrees. Seeds of most hardy plants—as wheat, oats, rye, lettuce, cabbage, and wild plants—germinate best in temperatures between 45° and 65°; those of tender vegetables and conservatory plants, between 60° and 80°; those of tropical plants, between 75° and 95°.

2b. The raising of seedlings

222. The ideal soil in which to plant seeds is loose and friable, does not "bake," and is retentive of moisture. It is neither hard clay nor loose sand.

223. The looser the soil, the deeper the seeds may be planted, since the plantlets can easily push through the earth; and the deeper the planting the more uniform is the moisture. For seeds of medium size and of strong germinating power,—as wheat, cabbage, apple,—a quarter or half inch is sufficient depth. In order to secure moisture about the seeds, the earth should be firmed or packed over them, particularly in a dry time; but this surface earth is moist because water is passing through it into the air (103, 104).

224. The smaller the seed, the shallower should it be sown, as a rule, and the greater should be the care in sowing. Very small seeds, as those of begonia, should be merely pressed into the earth, and the surface is then kept moist by shading, laying on a paper, cloth or glass, or by very careful watering. Delicate seeds are often sown on the surface of well-firmed soil, and are then lightly covered by sifting soil or dry moss over them. Keep them shaded until germination is well progressed.

225. Seeds may regerminate. That is, if germination is arrested by drought, the process may be renewed when congenial conditions recur, even though the young root may be dried and dead. This is true of wheat, oats, maize, pea, onion, buckwheat, and other seeds. Some seeds have been known to resume germination five and six times, even when the rootlet had grown half an inch or more and the seeds had been thoroughly dried after each regermination.

226. Bony and nut-like seeds must generally be softened by lying long in the earth; and the softening and splitting of the coverings is hastened by freezing. Such seeds are peach pits, walnuts, haws, and most tree seeds. Gardeners bury such seeds in earth in the fall, and plant them the following spring. The seeds are, also, often mixed with sand, or placed between layers of sand in a box, and if the seeds are from hardy plants the box of sand is placed where it will freeze throughout the winter. This operation is known as stratification.

3. *Propagation by Buds*

3a. *Why and how bud propagation is used*

227. When varieties do not "come true" or do not reproduce themselves from seeds, it is neces-

sary to propagate them by means of buds. In some cases, also, seeds are not produced freely, and then recourse is had to buds. In many instances, too, as in grafting, quicker results are obtained by bud propagation than by seed propagation. One means of dwarfing plants is to graft them on kinds of smaller stature.

228. Of bud propagation, there are two general types,— that in which the bud remains attached to the parent plant until it has taken root, and that in which the bud is at once separated from the parent plant. Examples of the former are layers; of the latter, cuttings.

3*b*. *Undetached buds*

229. A layer is a shoot or a root which, while still attached to the plant, is made to take root with the intention that it shall be severed, and form an independent plant.

230. The layers are bent to the ground, and at one place or joint are covered with earth; at this joint roots are emitted. Layering may be performed in either fall or spring, but the former is usually preferred. The layers are usually allowed to lie one season before they are severed. Almost any plant which has shoots that can be bent to the ground can be propagated by layers; but the best results are obtained in plants which have rather soft wood.

3c. Detached buds

231. Of propagation by detached buds, there are two types,—buds which are inserted in the soil or in water, and those which are inserted in another plant. The former are cuttings; the latter are grafts.

232. Cuttings may be made of soft or unripe wood, or of hard and fully matured wood. Of the soft kinds are cuttings (or "slips") of geraniums, fuchsias, and the like. Of the hard kinds are cuttings of grapes and currants.

233. Soft cuttings are made of shoots which are sufficiently mature to break or snap when bent double. They comprise at least one joint, and sometimes two or three. The leaves are removed from the lower end, and if the upper leaves are large they may be cut in two, or sheared, to prevent too rapid evaporation. A soil free from vegetable matter, as sand, is preferable. It is generally necessary to shade the cuttings until they are established.

234. Hardwood or dormant cuttings are taken in fall or winter. They usually comprise two or more buds. They root better if they are callused (partially healed over on the bottom end) before they are planted: therefore, it is customary to bury them in sand, or to stand them in sand, in a cool cellar until spring. In

spring they are set into the ground up to the top bud.

235. Single-eye cuttings—that is, one-bud cuttings—are sometimes employed when buds are scarce, as in new or rare plants. These are usually started under glass. They are planted half an inch or an inch deep, in an oblique or horizontal position.

236. Grafting is the operation of making one plant, or a part of it, grow upon another plant. The part which is transferred or transplanted is the cion; the plant into which this part is transplanted is the stock.

237. A cion may contain one bud or many. It may be inserted in a cleft or split in the wood of the stock, or it may be inserted between the bark and wood of the stock. A single bud which is inserted between the bark and wood is technically known as a "bud," and the process of inserting it is known as budding; but budding is only a special kind of grafting.

238. The cion and stock unite because the cambium of the two grow together. This cambium is between the bark and the wood (207): therefore it is important that the inner face of the bark of the cion (or bud) be applied to the surface of the wood of the stock; or, if the cion is inserted in a cleft, that the line between the bark, in the two, come together.

239. When the cion is inserted, the wounded surfaces must be tightly closed, to prevent the parts from drying out. Whenever the stock is cut off to receive the cion, thereby wounding the wood, wax is used to cover the wound; when only the bark is raised to admit the cion or bud, a bandage is used.

240. Grafting with hardwood cions of two or more buds — which is usually spoken of as grafting proper — is performed in spring, and the cions are cut in the winter and are kept fresh and dormant (as in a cellar) until wanted. The cion is made from the wood of the previous season's growth, of the variety which it is desired to propagate.

241. Budding—or inserting a single bud underneath the bark—may be performed whenever the bark of the stock will peel or "slip," and when mature buds can be secured. If performed in spring, the buds are cut in winter, as for grafting proper. If performed in late summer or early fall—and this is the custom—the buds are cut at the time, from the season's growth.

SUGGESTIONS ON CHAPTER IX

215*a*. It is impracticable, in this connection, to fully explain why it is that some plants "come true" from seed, and others (as apples, strawberries, roses) do not; but the enquirer will find the matter expounded in Bailey's "Plant-Breeding," pp.

88-91. The reason is that in plants which are habitually propagated by seeds, as the garden vegetables, we are constantly discarding the forms which do not come true, and are thereby fixing the tendency to come true,— since only the individuals which do come true are allowed to perpetuate themselves. In plants which are not habitually propagated by seeds, this selection does not take place, and the tendency to come true is not fixed.

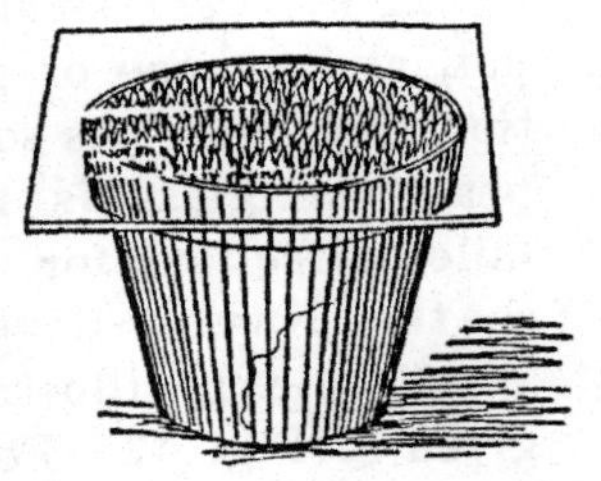

Fig. 46. Seed-pot, covered with glass.

217*a*. The longest-lived seeds are those borne on plants which reach their normal, healthy development. Those produced in very dry years are apt to have low vitality. Seeds should be stored in a dry and fairly cool room. Tables of the longevity of garden seeds may be found on pp. 104-107 of the 4th edition of "Horticulturist's Rule-Book."

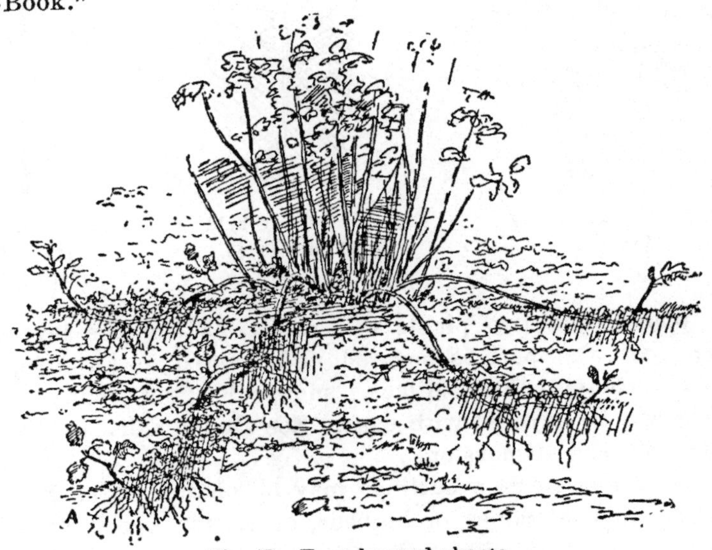

Fig. 47. Four layered shoots.

219*a*. "Nursery-Book," pp. 1-7, discusses the means of regulating moisture, with illustrations.

220*a*. As an experiment, plant corn a foot deep in warm,

firm soil. Run a little stick or splinter down to some of the seeds, allowing it to remain. The air enters alongside the stick. Observe if there is any difference in germination. If not, try it when the soil is very wet.

224*a*. Very small seeds are often sown very shallow in a pot, and a pane of glass is laid over the pot to check evaporation (Fig. 46). As soon as the plantlets appear, the glass is removed. For detailed directions for the sowing of seeds, see the "Nursery-Book," pp. 15–25.

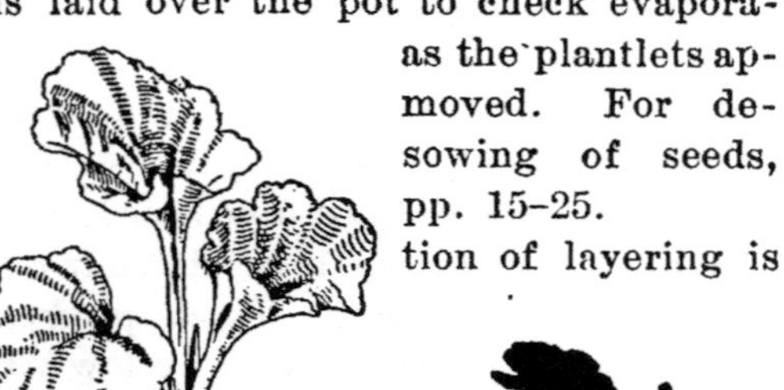

230*a*. An illustration of layering is given in Fig. 47. Four shoots are layered. One shoot, *A*, is layered in two places, and two plants will result. When the layers have taken root, the part is severed and treated as an independent plant. Honeysuckles, lilacs, snowballs, and many common bushes can be layered with ease. See Chapter iii., in "Nursery-Book," for full discussion.

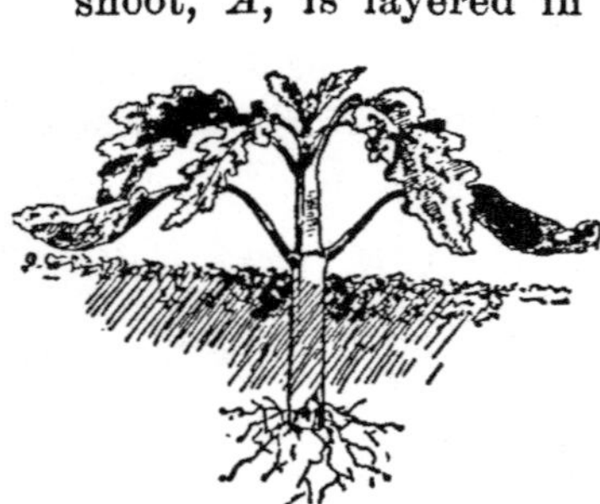

Fig. 48. Coleus cutting (x⅓).

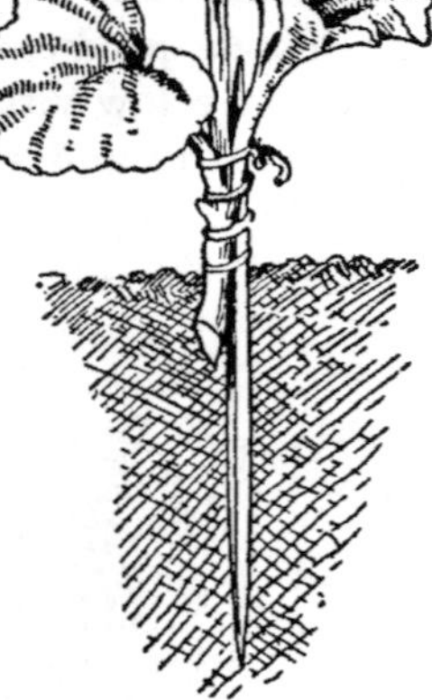

Fig. 49. Cutting held by tooth-pick (x⅓).

Fig. 50. One style of chrysanthemum cutting (x⅓).

233*a*. These green cuttings may be planted in shallow boxes of sand, in coldframes or hotbeds, or in the bench of a glasshouse. Figs. 48–50 illustrate the process.

234*a*. A grape cutting is shown in Fig. 51. This is the common fashion for propagating the grape; but new varieties are often grown from single eyes, as shown in Fig. 52. Consult

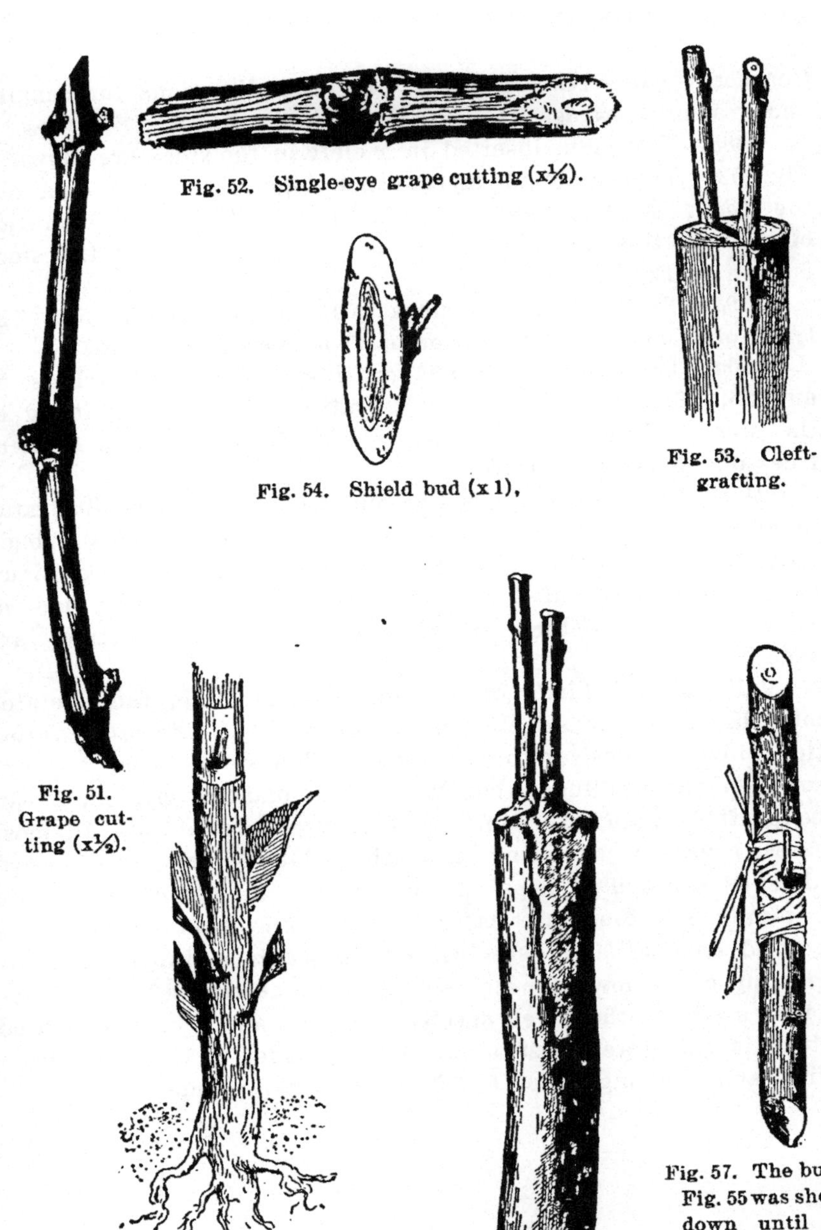

Fig. 52. Single-eye grape cutting (x½).

Fig. 53. Cleft-grafting.

Fig. 54. Shield bud (x 1),

Fig. 51. Grape cutting (x½).

Fig. 55. Bud entering matrix (x½).

Fig. 56. A waxed stub (x½).

Fig. 57. The bud in Fig. 55 was shoved down until covered by the bark, and now tied with bast.

Chapter iv. of "Nursery-Book," for full directions for making and growing cuttings.

238*a*. Two cions inserted in a cleft in the stock are shown in Fig. 53. The cambium layers come together in the cion and the stock. A "bud" cion is shown in Fig. 54, and the operation of shoving this down between the bark and wood of the stock is seen in Fig. 55.

239*a*. The waxing of a stock is illustrated in Fig. 56. The tying of a bud (by soft cord or bast) is shown in Fig. 57.

240*a*. The common style of grafting is suggested in Figs. 53 and 56. This is known as cleft-grafting, from the splitting of the stock. It is the style nearly always employed in orchard trees of apples and pears.

241*a*. Shield-budding is the common style. It is illustrated in Figs. 54, 55, 57. The buds are cut at the time of the budding, the leaves being at once taken off to prevent evaporation; but a bit of the leaf-stalk is usually left to serve as a handle, as seen in the picture. Peaches, cherries, plums, oranges, are usually budded.

241*b*. In all kinds of grafting and budding, the operator must be careful to select cions, or buds, from only those varieties which he desires to perpetuate. The stocks used by nurserymen are seedlings; but even if the plant is grafted, it can be grafted again, the same as if it were a seedling. In most cases, a variety is grafted on another plant of the same general kind, as a peach on a peach, an apple on an apple, a plum on a plum; but there are cases in which one kind or species is grafted on a different species: (*a*) to secure a dwarf plant, by grafting on a slow-growing root (as pear on quince), or (*b*) because seeds of the given species are rare, and a closely related stock is therefore substituted. For extended accounts of budding and grafting, refer to "Nursery-Book," Chapter v.

CHAPTER X

PREPARATION OF LAND FOR THE SEED

I. P. ROBERTS

1. *Factors Which Determine the Preparation of the Seed-bed*

242. Faulty preparation of the land is the cause of more failures than the subsequent treatment of the crop. In field conditions, this preparation can not be so thorough, or so ideal, as in garden areas or in glass-houses. The general condition of the farm work dictates to a great extent the particular time when the seed shall be sown and the amount of preparatory work which shall be put on the land: therefore, it is very important that the farmer fully understand what is required, in order that he may make no mistakes.

243. The preparation of the land for seeding should be governed by two factors: by the needs of the particular plant which is to be grown, and by the character of the land. To prepare a seed-bed for any crop, the habits, likes and dislikes of the plants should be

studied. That is, it is not enough that the land be well prepared: it should have the kind of preparation which is demanded by the crop.

2. *The Demands of the Plant*

244. The preparation of the seed-bed differs with the way in which the plant is propagated. Some plants are propagated by a piece or part of an underground stem or tuber, as the potato; others by a branch of the aerial part, as the willow or sugar-cane. In all of these cases, the buds or eyes are surrounded with food for immediate use. This stored food gives them power to send out strong shoots and to grow for some time without having to secure moisture from the soil. But many plants are propagated by tiny seeds. These start in life with little stored food, and, therefore, must quickly secure nourishment from the soil; and the land must, therefore, be very well prepared. These seeds should be planted near the surface, for there will not be strength enough in the infant plant to push its way through, if planted as deep as the potato.

245. Plants may change or modify their characteristics to adapt themselves to changed conditions. The common red clover is a tap-rooted plant, but if it grows on soil which is

underlaid with wet clay, it tends to become fibrous-rooted. Even long-lived perennials, as trees, do best when the surface soil is well prepared to a depth of ten to twelve inches, since many feeding roots of trees, especially of young ones, find nourishment in this prepared soil.

246. Plants differ greatly, however, in ability to adapt themselves to unfavorable conditions. Many common plants send their tap-roots into the subsoil for two to three feet, even if it be hard, while sugar beets become fibrous-rooted, and may be pushed up and partly out of the ground if their tap-roots attempt to enter the undisturbed hard subsoil. Land devoted to clover need not necessarily be subsoiled if it be fairly free from stagnant water, while that planted to sugar beets should be subsoiled, for the reason that a long, fusiform root is desired, all or nearly all of which should be below the surface; for that part of the beet which grows above the ground is not nearly so valuable for making sugar as that part which grows under ground.

247. Nearly all of the common and quick-growing plants secure the larger part of their nourishment and moisture from the first, or surface foot of soil. This being so, it is seen how necessary it is to prepare the soil in the best possible manner. If the upper soil is not well

prepared, the roots must search wide and deep for food.

248. Most of the smaller plants require but about six months in which to grow and to fruit. If, in order to secure nourishment and moisture, the roots are obliged to descend into the cold, hard subsoil, where the plant-food is likely to be least available, neither growth nor fruitage can be satisfactory. Those plants which do not mature until they are five to twenty years of age, as fruit trees, can secure much nourishment from the subsoil, although they secure little in any one growing season. Then, too, trees must secure a firm hold on the land, or they will be prostrated by winds. By being obliged to send many of their roots into the cold, firm subsoil through many generations, trees have probably acquired the power of securing more of the tough or unavailable food of the subsoil than plants which live but one season.

249. Different plants require not only to be planted at different seasons of the year, but at different depths. They demand different methods of preparation of the surface soil. Some do best when placed in loose, warm soil, as, for instance, maize and sweet potatoes; while others do best when grown on fairly cold and somewhat compacted surface soil, as winter wheat.

3. *The Preparing of the Seed-bed*

250. Nearly all plants thrive best when furnished with a full and continuous supply of moisture. Fine, loose earth, which contains a moderate admixture of humus, is capable of holding much moisture (73, 74); but the soil may be so loose and light as to admit too rapid movement of air, in which case the moisture will be carried away. If the particles of earth are separated too widely, capillarity is weakened. In such cases the subsurface soil should be slightly compacted, while one to three inches of the surface is left loose to form an earth-mulch, which tends to prevent loss of moisture by evaporation. The particles of the loose surface earth-mulch should be so widely separated that the moisture can climb only to the bottom of it, for if it comes to the surface the air will carry it away (83). The earth-mulch shades the ground in which the plants are growing, prevents the soil from cracking, and saves moisture.

251. The seed-bed should contain no free water; but it is impossible to secure this condition at all times. No serious harm will come when the soil is over-saturated at planting time, if the free water is quickly removed. If the soil contains more water than it can hold by

capillarity, the air is driven out, and the soil swells and tends to become puddled (81).

252. Many seeds will not germinate if planted out of season, or when the soil is cool, no matter how well the seed-bed is prepared. Then, if it is desired to plant early, make the land fine and loose, for in so doing the temperature of the soil is raised. The soil of a fine, porous seed-bed, resting on a well-drained subsurface and subsoil, is much warmer than one resting on a compact, undrained foundation. However, it is not wise to plant seeds out of season or when the weather is unsuitable.

253. If small seeds are covered with but little earth, they may fail to germinate for lack of moisture. If covered with enough fine earth to insure a constant supply of moisture, the young plants have a hard struggle to reach the surface. Only a few of the small seeds, as clover and many of those planted in the kitchen-garden or flower-garden, ever produce plants. Sometimes the seeds are imperfect, but more often the failure to secure vigorous germination is due to a poor seed-bed or to careless planting. To obtain better results, not only prepare a fine seed-bed and sow at the proper time, but compact the soil immediately over the row of seeds. This will enable capillary attraction to bring moisture to the surface, or near it (103). The

earth-mulch should remain unpacked between the rows, to conserve moisture.

254. In some cases it is impossible to secure a proper seed-bed for small seeds. For example, no suitable seed-bed can be procured, as a rule, for clover seeds when sowed in a growing tilled crop. In order to secure germination, these seeds are sown on the surface in early spring, while the surface soil is still porous from winter freezing. The spring rains wash the seeds into the little cracks in the soil and partly cover them. The weather being cool and cloudy and the soil moist in early spring, the oily seeds of the clover are kept damp enough to insure germination. If such small seeds are sown in summer or early fall, the land is rolled for the purpose of supplying them with moisture.

255. A good field seed-bed, then, can be secured profitably only on land which is either naturally or artificially well drained, which has been well broken and crumbled by the plow, and the surface of which has been thoroughly fined by the harrow. Particular care should be taken not to work heavy or clay lands when they are wet. Neither should clay lands be tilled so much that they become very dusty, else they will puddle when the rains come. The remarks respecting the proper tillage of the land (Chapter iv.) will apply here.

4. *Application of the Foregoing Principles*

4a. *Wheat*

256. Winter wheat does best when one or two inches of the surface soil is fine and loose, and the subsurface soil fine and fairly compact.

257. To secure the ideal conditions, the ground should be plowed some time before sowing, and the manure spread on the rough surface. The ground is immediately harrowed, rolled, and harrowed again. In one or two weeks afterward it is surface-tilled again, with the implements best suited to the particular soil. All this tends to divide and cover the manure, compact the subsurface soil, form a fine seed-bed, conserve moisture, and set free plant-food.

258. This treatment of the land causes the roots to be many and fibrous, and to remain near the surface, where the plant-food is most abundant and available. If the manure is plowed under and the soil remains loose, the roots are less fibrous and descend to the bottom of the furrow. In the spring, it often freezes at night and thaws during the day. This tends to lift the plants and to break their roots. But if the roots are nearly horizontal and near the surface, they tend to rise

and fall with the freezing and thawing, and are not seriously injured.

259. As the soil becomes hot at the surface in June and July, the shallow roots descend to the subsurface soil, where it is cool and where the plant-food was not drawn upon during the fall; while the deep fall-rooted plants will be unable to find new feeding ground when they need it most, just before fruiting, unless the roots start toward the surface, which they will not do, for in midsummer the surface soil is hard and dryish and too warm for wheat roots.

4b. Maize, or Indian corn

260. The seed-bed for maize, which is a sun-plant and does best when planted in a warm soil, may be prepared in a different way from that designed for winter wheat. Since maize is planted in the spring, when the soil is often too cool for this semi-tropical plant, the subsurface soil should not be as compact as for wheat. If left rather open, the warm spring rains pass quickly to the subsoil and warm the soil (77). The more open seed-bed will allow a freer circulation of warm air through the soil.

261. The best machines for planting maize are those which deposit the seed one to two inches below the surface in the fine, moist soil,

and compact the surface soil over the seed by means of concave wheels about eight inches wide, while the spaces between the rows are not compacted. The maize may be cultivated and harrowed before the plants appear, since the rows may be easily followed by the marks left by the concave roller wheels. The frequent inter-tillage which will be required to destroy weeds, to preserve the earth-mulch, and to set free plant-food, will compact the subsurface soil quite as much as is desirable.

4*c*. *Potatoes*

262. The potato should be planted deep and left with uncompacted surface soil. The seed potato contains about 75 per cent of moisture, and has a large quantity of stored food for nourishing the buds and sending up strong shoots. It thrives best in a cool, moist soil; and this condition is secured if it is planted about four inches deep.

263. It should also be remembered that potatoes are enlarged underground branches, and that the new tubers prefer to grow above the seed-tuber. If the seed-tuber be planted shallow, the branch or stem above the seed is so short that there is little room for underground stems.

264. Usually potatoes should not be hilled at

the last cultivation, for at that time the potatoes will have begun to form near the surface or in the subsurface soil, according to soil conditions, moisture, climate and variety. Then, to throw a mass of dirt on top of these underground stems, after they have chosen the best position for highest development, is to force them to adapt themselves to new conditions.

SUGGESTIONS ON CHAPTER X

242*a*. In this chapter, the word seed is used in its general agricultural sense, to designate seeds or other parts (as tubers) which are planted for field crops.

243*a*. A seed-bed is the soil in which the seed is planted or sown. It may be the size of a window box, a hotbed frame, a garden bed, or a field of wheat.

244*a*. The sprouts which appear on potatoes in cellars are supplied from the nutriment stored in the tuber. If a winter branch of a tree is stood in water in a warm room, leaves and sometimes flowers will appear in the course of a few weeks ; and the growth is made from the nutriment stored in the twig. All seeds have stored nutriment, but the small ones have very little, and it may be exhausted before the plantlets can get a foothold in the soil. The better and finer the seed-bed, the sooner the plantlet can establish itself.

250*a*. The subsurface soil is that lying just below the surface, —between the surface and the subsoil. It is the lower part of the soil which has been loosened by the plow,—that part which is below the reach of the surface tilling.

250*b*. The subsurface soil may be compacted by rolling (102), after which the surface is loosened by harrowing. When land is given much surface tillage, as for wheat, the tramping of the horses compacts the under soil. Loose, sandy lands may be plowed shallow in order to keep the subsurface compact (94).

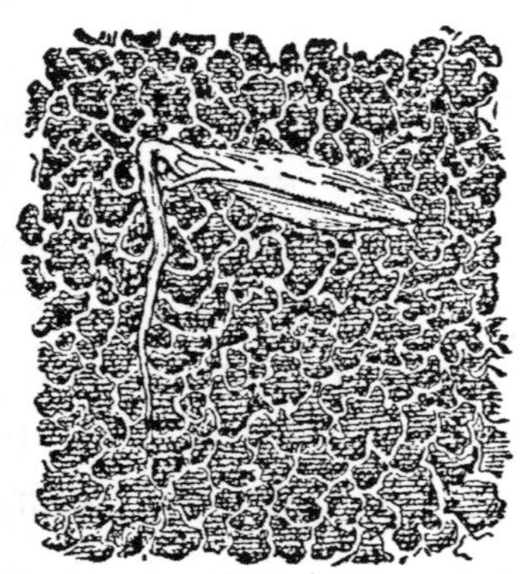

Fig 58. A well drained but moist soil.

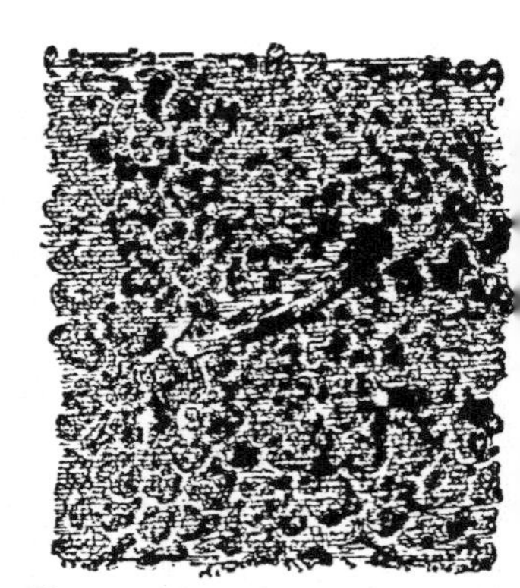

Fig. 59. A wet and uncongenia soil.

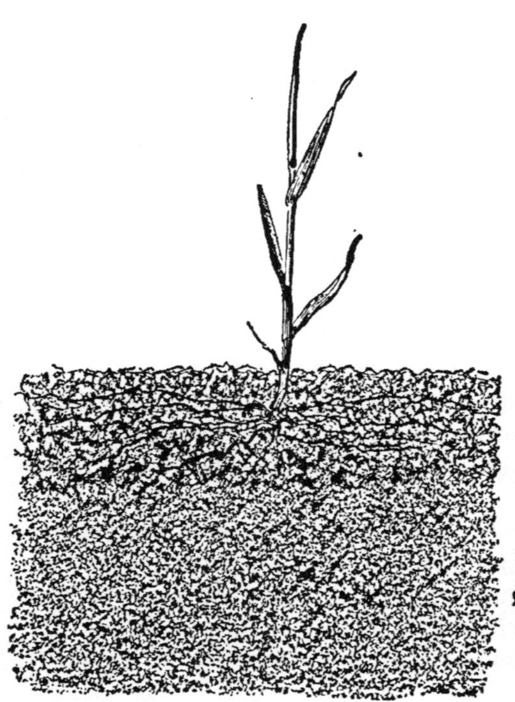

Fig. 60. A wheat plant properly grown, in the fall.

Fig, 61. The result of too loose so and manure plowed under.

251*a*. The Fig. 58 shows a drained soil supplied with moisture held by capillarity in the smaller interstices, while the larger channels have been relieved of free water by percolation. Fig. 59 represents a supersaturated soil from which air and heat are largely excluded. If seeds remain for a few days in this undrained soil they fail to germinate, and may rot. Should stagnant water remain in the soil for some time after the plants have appeared above ground, they will turn yellow, and may perish (194). All this emphasizes the necessity of preparing a seed-bed adapted to the wants of the plant to be grown, and of maintaining such soil conditions as are best suited to the wants of the plant during its entire period of growth.

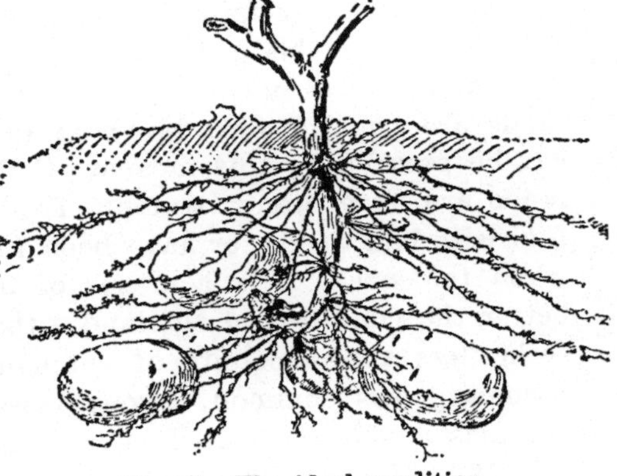

Fig. 62. The ideal condition.

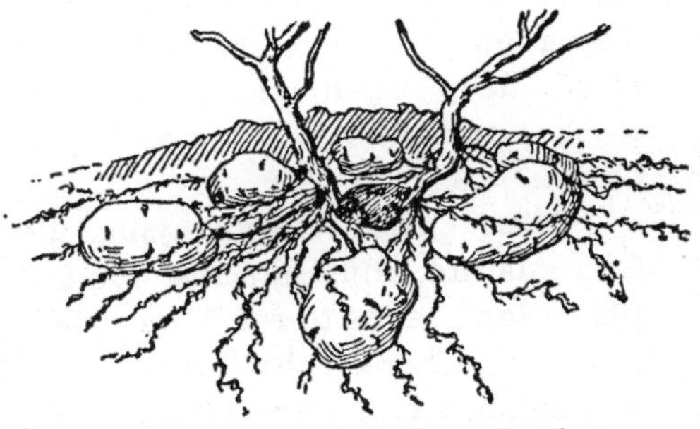

Fig. 63. The result of shallow planting.

253*a*. "Care should be exercised not to sow very small and slow-germinating seeds, as celery, carrot, onion, in poorly prepared soil or in land which bakes. With such seeds it is well to sow seeds of radish or turnip, for these germinate quickly and break the crust, and also mark the row, so that tillage may be begun before the regular-crop seeds are up."—*Bailey, Garden-Making, p.. 37.*

255*a*. The expense of preparing the land can often be materially diminished if the land is plowed some little time before it is planted, in such a way that the elements can act upon the soil through the process of weathering. In such cases, the furrow-slice is not laid flat, but left at an angle of about forty-five degrees, that the soil may become warmed for the purpose of promoting chemical action and the liberation of plant-food. It may also serve to hasten the drying of the land (95).

255*b*. Summer-fallowing is often an advisable means of preparing the seed-bed. It consists of two or more summer plowings and several harrowings, the land remaining idle. Fallowed lands are usually sown to wheat in the fall. An ideal seed-bed can be secured by this means. Fallowing is to be advised when lands are very stony, stumpy, hard, or when they have become foul with bad weeds, or have been injured by plowing or ditching when too wet. It is a means of putting the land right. The better the condition of the land,—that is, the better the farming, —the less the necessity of summer-fallowing. The practice is becoming less common, largely because modern implements and methods enable us to handle the land better.

258*a*. The pictures will make this reasoning plain. Fig. 60 represents a wheat plant in the fall, on properly handled land. The roots are near the surface. Fig. 61 shows how the roots strike deep when manure is plowed under and the soil is left loose; and this plant stands less chances of success than the other.

263*a*. The accompanying figures, which are made directly from nature, illustrate the point that deep planting in well-prepared land tends to result in a deep and spreading hill of potatoes (Fig. 62), whereas shallow planting in poorly prepared land results in a shallow and crowded hill (Fig. 63). The better potatoes may be expected in the former case.

CHAPTER XI

SUBSEQUENT CARE OF THE PLANT

1. *By Means of Tillage*

1*a*. *In general*

265. Tillage is the first consideration in the care of the plant. This is emphatically true in the field; but in the glass-house tillage is reduced to a minimum, in part because the preparation of the soil is so thorough.

266. The objects of tillage, in the care of the plant subsequent to seeding or planting, are three: (*a*) to supply plant-food, by rendering the soil constituents available; (*b*) to supply moisture; (*c*) to destroy weeds. The first two captions have been discussed in Chapters ii., iii., iv.

267. (*c*) Weeds are only incidental difficulties. They are the results of faulty management of the land. If the first attention is given to the crops and the land, the question of weeds will largely take care of itself. It is less important to know the kinds of weeds than it is to know how to till and to crop the land.

268. There are four general means of keeping weeds in check: (*a*) by good tillage (101, 101*a*); (*b*) by rotation of crops, by means of which any one kind of weed is prevented from becoming thoroughly established; (*c*) by complete occupation of the land with crops,—for weeds find opportunity when the ground is not fully occupied, as in old and thin meadows; (*d*) by killing the weeds directly.

269. Surface tillage should be given as often as the ground becomes hard, or whenever the earth-mulch needs repairing (100). Under general conditions, tilled crops, as maize and potatoes, should be cultivated every ten days or two weeks, particularly early in the season. As soon as low crops cover the ground, and thereby afford a mulch, cultivation may cease.

270. Sowed crops can often be tilled once or twice to advantage very early in the season, by running a fine-toothed harrow over them. Thus, wheat and maize are now often harrowed in early spring. The harrowing destroys but few plants, while it loosens the soil, and conserves moisture before much has been lost by hot weather. Harrowing meadows and pastures causes the plants to tiller or to stool out, and thereby to cover the ground more completely; it also breaks the old, hard roots and causes new feeders to appear, thereby re-invigorating the plants.

1b. *In fruit plantations*

271. Tillage gives the same results in fruit plantations as with annual crops, and it also has particular advantages in such cases: it causes the roots of the trees or bushes to strike deep into the soil and thereby to find moisture in dry times, and it has a decided effect in keeping down the ravages of insects and the incursions of diseases by destroying breeding-places and burying diseased foliage and fruit.

272. Since fruit trees and bushes send their roots so deep into the soil, they are better able to withstand neglect of tillage than annual crops are. There has thus arisen a general belief that orchards do best in sod; but in most cases of successful sod orchards the trees thrive in spite of the sod, not because of it.

273. It is particularly important to till fruit plantations early in their life. Apples should generally be tilled for at least the first ten years. The plants thereby get a good start and come into bearing early; and the habit acquired in the first years is apt to continue. The treatment given in the early period usually determines the success of the fruit plantation.

274. The fruit plantation may need tillage throughout all the years of its existence, and, as a matter of fact, it usually does need it. But if

the trees or bushes tend to grow too fast, so that they do not bear, or become top-heavy, or do not stand the winter, they may be checked by putting the plantation in sod; but even then, the sod is only a temporary expedient. If the management of the plantation has been right, it is doubtful if sod can ever be an advantage,—or at least with none of the common fruits, except possibly apples and pears.

275. All fruit plants start into growth very early in the season. Therefore, tillage should be begun the moment the ground is fit; and it should be continued unremittingly until the time arrives for all tillage to cease.

276. The growth on fruit plants generally ceases by midsummer. Therefore, tillage may stop at midseason or early fall; and at the last tillage a cover-crop may be sown (109, 114, 116). Stopping the tillage early allows the plants to mature their growth, and thereby be more likely to escape winter injury; and it lessens the danger of overgrowth. If the trees are carrying a heavy crop, however, it may be necessary to continue the tillage in order to supply the fruit with moisture, especially if the land or the season is dry.

277. The tillage of fruit-plantations usually consists of a spring plowing, followed by harrowing. If the land has been well handled in

the first few years, deep and heavy plowing will not be needed when an orchard comes to maturity. Light gang-plows, or even cultivators, may then be sufficient for the first breaking of the soil in spring.

2. *By Means of Pruning and Training*

2*a*. *Pruning vs. training*

278. Pruning is the removing of certain parts of plants for the purpose of augmenting the welfare of the plant or to secure more, larger or better products (as better fruit or flowers). Training is the trimming or shaping of the plant into some particular or desired form. Successful pruning depends upon principles of plant growth; training depends upon the personal ideal of the pruner.

279. Nature prunes. In every plant, more branches start than can ever mature; and many buds are suppressed before they have made branches. Every tree top, if left to itself, will sooner or later contain many dead branches. There is a struggle for existence amongst the branches, and the weakest die.

2*b*. *The healing of wounds*

280. Pruning depends upon two sets of factors,—upon the questions concerned in the heal-

ing of wounds and the injury to the plant, and upon the general results which it is desired to attain. Knowing how wounds affect the plant, the pruner should then have a definite purpose in view when he cuts a limb.

281. The proper healing of wounds depends primarily upon (*a*) the kind of plant (observe that peach trees heal less readily than apples), (*b*) the vigor of the plant, (*c*) the position of the wound on the plant (wounds on strong main limbs heal better than those on weak or side limbs), (*d*) the length of the stump—the shorter the stump the quicker the healing,—(*e*) the character of the wound as to smoothness or roughness.

282. Other matters which determine the proper healing of a large wound are (*f*) the healthfulness of the wood, (*g*) the season of the year in which the cut is made, (*h*) the protection which the wound receives.

283. (*g*) Other things being the same, wounds heal quicker when made in the early part of the growing season,—that is, in late spring; but the factors mentioned in 281 are more important than the season.

284. (*h*) Dressings do not, of themselves, hasten the healing of wounds, but they may keep the wound sound and healthy until it heals of itself. A good dressing is one which is anti-

septic and durable, which affords mechanical protection, and which does not of itself injure the tissue of the plant.

2c. *The principles of pruning*

285. We prune (*a*) to modify the vigor of the plant, (*b*) to produce larger and better fruits or flowers, (*c*) to keep the plant within manageable shape and limits, (*d*) to make the plant bear more or bear less, (*e*) to remove superfluous or injured parts, (*f*) to facilitate spraying and harvesting, (*g*) to facilitate tillage, (*h*) to make the plant assume some desired form (properly, training).

286. Heavy pruning of the top tends to increase growth, or the production of wood. Heavy pruning of the root tends to lessen the production of wood. Water-sprouts generally follow heavy pruning, particularly if the pruning is performed in winter.

287. Checking growth, so long as the plant remains healthy, tends to cause overgrown plants to bear. One means of checking growth is to withhold fertilizers and tillage; another is to resort to root-pruning; another is to head-in or cut-back the young shoots. Some plants, however, bear most profusely when they are very vigorous; but they are such, for the most part, as have been moderately and continuously vig-

orous from the beginning, rather than those which are forced into very heavy growth after a long period of neglect.

288. The heading-in of young growths tends to force out the side shoots and to develop the dormant buds. The more a plant is headed-in, therefore, the more thinning-out it will require. Heading-in induces fruitfulness by checking growth and by encouraging the formation of side spurs (upon which fruit may be borne).

289. Heavy pruning every few years—which is the custom—tends to keep trees over-vigorous and unproductive. Mild pruning every year maintains the equilibrium of the plant, and tends to make it fruitful.

3. *By Keeping Enemies in Check*

3a. *The kinds of enemies*

290. Of plant enemies or diseases, there are three main types,—insects, parasitic fungi, constitutional or physiological troubles.

291. Insect pests are of two general types, so far as their method of feeding is concerned,—insects which chew, or bite off pieces of the plant, and those which suck their food from the juices of the plant. In the former class are the worms and beetles; in the latter are plant-lice, scale insects, and the so-called true bugs (as the

squash-bug or stink-bug, and the leaf-hoppers). We may classify injurious insects again, without reference to their mode of taking food, into those which live and feed on the outside of the plant, and those which, as borers and apple-worms, burrow and feed inside the tissue.

292. Of fungous pests, the farmer may recognize two groups,—those which live wholly on the outside of the host (as the powdery mildew of the grape, pea mildew), and those which live wholly or in part inside the tissues (as apple-scab, black-knot, potato mildew). Most injurious fungi are of the latter kind. Fungous troubles are nearly always marked by definitely diseased spots on the leaves or twigs.

293. Physiological or constitutional troubles are those which affect the whole plant or an entire leaf or branch, and the cause of which is not apparent on the exterior. These troubles may be due to germs or bacteria working within the tissues (as pear-blight), or to some difficulty in the nutrition of the plant. These troubles are generally not marked by definitely diseased spots or blemishes, but by the gradual dying of an entire leaf, branch or plant.

3b. The preventives and remedies

294. Keeping the plants vigorous and healthy is the first step towards the control of pests and

diseases. Clean tillage, rotation of crops, planting varieties which are least liable to attack, and careful attention to prevent all the conditions which seem to favor the breeding of insects and the spread of diseases, are quite as important as destroying the enemies; for "an ounce of prevention is worth a pound of cure."

295. Insects are destroyed by three general means: (*a*) by killing them directly, as by hand-picking, digging out borers; (*b*) by killing them by means of some caustic application to their bodies; (*c*) by poisoning them by poisoning their food. In some instances, insects may be kept away by covering the plants with some material, as lime, to which the insects object; but this method of fighting insects is usually unsatisfactory. A substance which is used to destroy an insect is called an insecticide.

296. (*b*) The caustic applications or insecticides must be used for those insects which suck their food (291). Kerosene, kerosene emulsion, soap washes, tobacco, and the like, are the materials used; and plant-lice, scale insects, plant-bugs, thrips, and leaf-hoppers are the insects thus treated.

297. (*c*) The poisonous applications are used for the chewing insects which prey upon the outside of the plant (not for borers, which are usually dug out). Paris green, London purple,

and white hellebore are the materials commonly used; and worms, potato-bugs, and all leaf-chewing pests, are the insects thus treated.

298. Fungi are killed by materials which contain sulfur or copper. Fungi which live inside the leaf or stem (292) cannot be killed directly by applications, but the parts which project into the air (the fruiting portions) can be destroyed and the fungus thereby weakened and checked; and the spores (which answer to seeds) cannot grow on a surface which is covered with copper or sulfur. The best treatment of plant diseases, therefore, is to make the application before the disease gains a foothold. A substance which is used to destroy fungi is called a fungicide.

299. The best general fungicide is the Bordeaux mixture, made of lime and sulfate of copper. It not only destroys the fungi, but adheres long to the plant. Another good fungicide is carbonate of copper; and it is preferred for ornamental plants and for late application to fruit, because it does not discolor or soil the leaves or fruits.

300. The application of insecticides and fungicides is usually made in water, with a syringe or pump, or by means of a spray; and thereby has arisen the practice of spraying.

301. In order that spraying shall be successful, it must (*a*) apply the materials which will

destroy the pest in question and yet not injure the plant, (*b*) be thoroughly done, so that no part of the plant is left unprotected, (*c*) be performed the moment the enemy appears, or, in the case of fungous diseases, as soon as there is reason to believe that the pest is coming.

302. The best machine or pump is the one which throws the finest spray the farthest distance. Other factors are the capacity of the pump, its strength, its durability, its lightness, the ease with which it works.

303. Spraying will not keep all fungous diseases in check; and, in any case, it should be supplemented by sanitation, as by burning or burying the fallen diseased leaves and fruits, the cutting away of infected parts, and the like. Some fungous diseases, as the grain smuts, are carried over from year to year in the seed; and the proper treatment is to soak the seed in a fungicide. The constitutional diseases (293) must be treated by other means than spraying, usually by burning the affected part or plant (294, 294*a*).

SUGGESTIONS ON CHAPTER XI

267*a*. "The daisy-cursed meadows of the East are those which have been long mown and are badly 'run,' or else those which were not properly made, and the grass obtained but a poor start. The farmer may say that the daisies have 'run out' the grass, but the fact is that the meadow began to fail, and the daisies quickly seized upon the opportunity to gain a foothold. * * * The weedy lawns are those which have a thin turf, and the best treatment is to scratch the ground lightly with an iron-toothed rake, apply fertilizer, and sow more seed." "The agricultural conditions in the Dakotas and other parts of our Plains region are just such as to encourage a hardy intruder like the Russian thistle. An average of eight or nine bushels of wheat per acre is itself proof of superficial farming;

Fig. 64. A gang-plow.

Fig. 65. A light gang-plow for very shallow work.

but the chief fault with this western agriculture is the continuous cropping with one crop,— wheat."—*Bailey, "Survival of the Unlike," pp. 196, 195.*

270*a*. Maize may be harrowed until it is four inches high. The plants will straighten up. This harrowing is cheaper than cultivating; and if the land is put in good condition very early in the life of the crop, much less subsequent tillage is required. In general, narrow-toothed harrows should be used (Fig. 24), but the style of tool must be adapted to the particular land in question.

277*a*. If the plowing has been thorough for the first few years after the orchard is planted, the ground should be so mellow that very light plowing will answer thereafter. There will be no sod to tear up and to plow under, and the tree roots will be deep in the ground, where they can find moisture. A gang-plow (Fig. 64) should be sufficient for the spring plowing in most mature orchards, unless there is a heavy growth of cover-crop to plow under. A tool for still shallower plowing is shown in Fig. 65. This is excellent for orchards on light or loose soils, although its height makes it more difficult to handle

Fig. 66. The proper way to make the wound.

Fig. 67. The wrong way to make the cut.

about low-headed trees. For full discussions of the tilling of fruit plantations, see "Principles of Fruit-Growing," Chapter iii.

278*a*. If some of the limbs are taken from an apple tree for the purpose of making it bear better, the operation is pruning; if the tree is sheared or trimmed to make it round-headed, the operation is training. A rose or a grape-vine may be pruned by cutting away part of the wood; it may be trained on wires or to the side of a house.

279*a*. On the subject of the struggle for existence in the tree top, consult, Observation iv. in "Lessons with Plants," and Chapter i. in "Pruning-Book." The philosophical bearings of this fact of competition are presented in Essay iii., "Survival of the Unlike."

281*a*. Other things being equal, the closer the wound to the branch, the quicker it will heal. The smoother the wound, the better and quicker it will heal. Figs. 66 and 67 illustrate right and wrong methods. For full discussion of the healing of wounds, read Chapter iii. in the "Pruning-Book."

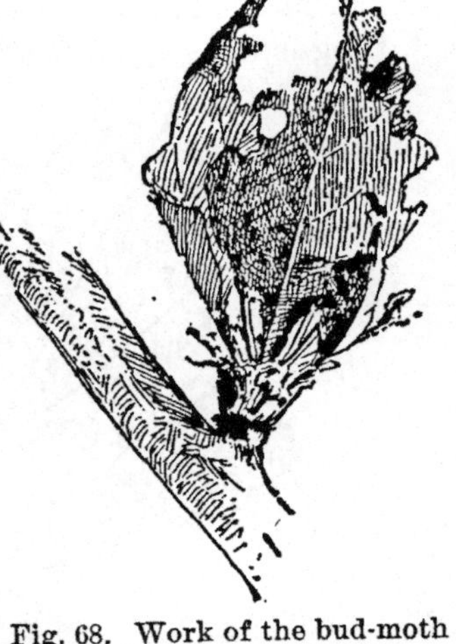

Fig. 68. Work of the bud-moth larva,—a chewing insect.

284*a*. An antiseptic dressing is one which prevents germs or microbes from growing on the surface of the wound; for the decay which follows wounds is the work of germs and fungi. In general, the best dressing for wounds is lead paint. Wax is not durable enough, nor is it antiseptic. Bordeaux mixture is good for its antiseptic properties, but is not durable, and it affords little protection from the weather.

285*a*. The principles of pruning are discussed under twenty heads in Chapter iv. of "Pruning-Book."

291*a*. The chewing or biting insects eat up the parts upon which they prey. Fig. 68 is an example of such work. The sucking insects do not eat up the part, but they often leave distinct marks of their work, as in Fig. 69. A plant-bug is shown in Fig. 70. The true weevils and curculios are biting insects, although they have snouts (Fig. 71).

292*a*. A fungus is a plant. It is destitute of chlorophyll or leaf-green. It lives on living organisms (or is parasitic), or on dead or decaying matter (or is saprophytic, as mushrooms and toadstools). Some kinds, as toadstools, are large and conspicuous; others, as molds, are small and fragile; while still others are nearly or quite microscopic. The plural of fungus is fungi (rarely written funguses). As an adjective, the word is written fungous, as a fungous disease. A fungoid disease is a fungus-like disease, the exact origin of which may not be known or specified. Rusts, mildews and leaf-blights are types of fungous diseases.

292*b*. The plant or the animal upon or in which a parasitic fungus lives is known as its host. The fungus injures its host by

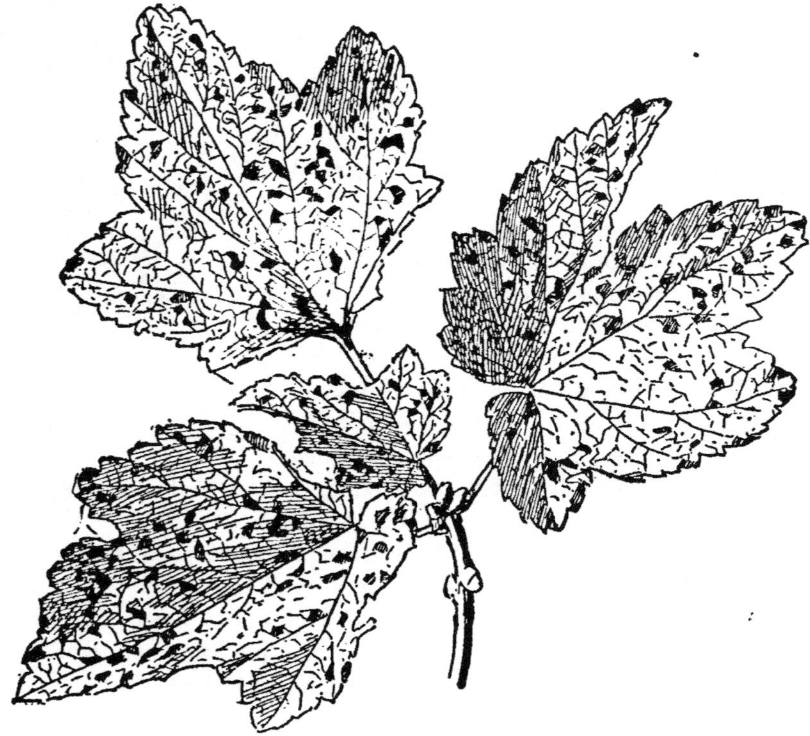

Fig. 69. Work of the four-lined leaf-bug—a sucking insect—on currant foliage.

robbing it of nutriment and sometimes by breaking up its cellular structure, and by obstructing the breathing-pores and interfering with the movement of its fluids.

293*b*. Physiological troubles may be termed internal troubles, although the germs which cause some of them enter from the outside. There is no external growth of a fungus, and rarely any well defined small spots on the leaves. Fig. 72 shows the spots of a fungous disease ; if this leaf had been attacked by a bacterial or physiological disease, the entire leaf would probably have shown signs of failing, for the food supply is usually cut off in the leaf-stalk or the main veins. In Fig. 72, however, each spot represents a distinct attack of the fungus. Fig. 73 is a type of physiologial trouble, the edge of the leaf dying from the cutting-off of its food supply ; this dead border will widen until the leaf dies.

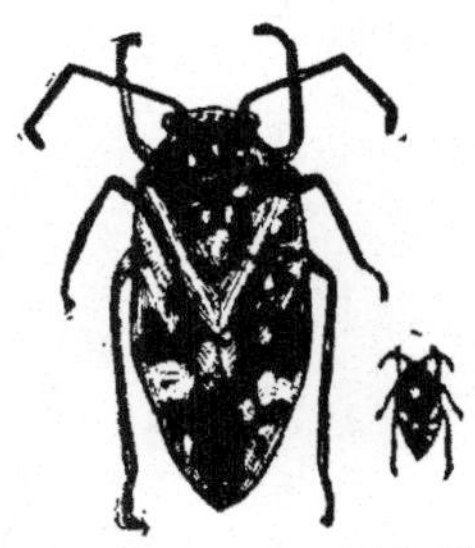

Fig. 70. The tarnished plant-bug,—a sucking insect.

Fig. 71. The strawberry weevil, — a chewing insect.

294*a*. Physicians treat some diseases by prophylaxis,—that is, by giving attention to means of sanitation and of preventing the spread of the disorder. Farmers must do the same. Wire-worms are rarely troublesome in short and quick rotations, particularly in those in which sod is not a prominent feature. Club-root of the cabbage is rarely troublesome on land which has not grown cabbages or allied plants for a few years. Apple-scab is least serious in those orchards which have been thoroughly sprayed in previous years. Plum-rot is least troublesome when the fruit is well thinned. Rose-bugs seldom trouble vineyards which are on strong or heavy lands.

296*a*. Kerosene emulsion is made as follows: Hard soap, ½-pound; boiling water, 1 gallon; kerosene, 2 gallons. Dissolve the soap in the water, add the kerosene, and churn with a pump for 5 to 10 minutes. Dilute 4 to 25 times before applying. Use strong emulsion for all scale insects. For such insects as plant lice, mealy bugs, red spider, thrips, weaker preparations will prove effective. Cabbage worms, currant worms, and all insects which have soft bodies, can also be successfully treated. It is advisable to make the emulsion shortly before it is used.

296*b*. Mixtures of kerosene and water are effective insecticides, and there are now pump attachments for mechanically mixing the two. One part kerosene to four parts water will kill nearly all insects and not injure the foliage. This may be expected to take the place of the kerosene and soap emulsion for most purposes.

297*a*. The Paris green mixture is compounded by using Paris green 1 pound, water 150 to 300 gallons. If this mixture is to

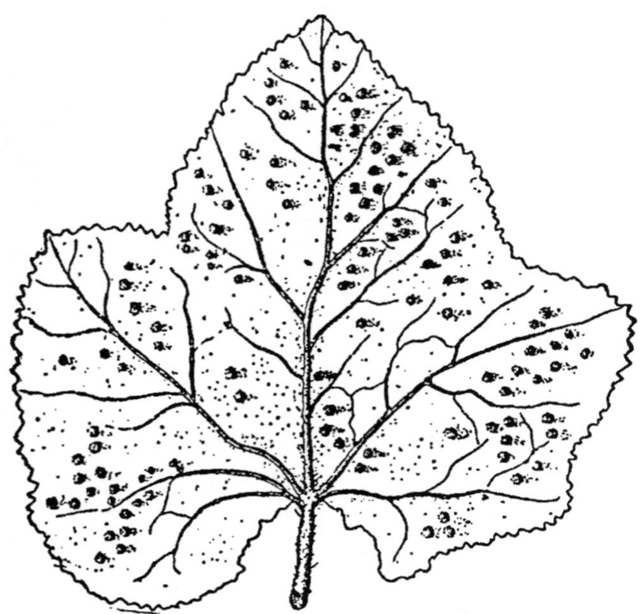

Fig. 72. The spots of hollyhock rust,—a fungous disease.

be used upon fruit trees, 1 pound of quicklime should be added. Repeated applications will injure most foliage, unless the lime is used.

297*b*. London purple is used in the same proportions as Paris green, but as it is more caustic it should be applied with two or three times its weight of lime, or with the Bordeaux mixture. The composition of London purple is variable, and unless good reasons exist for supposing that it contains as much arsenic as Paris green, use the latter poison. Do not use London purple on peach or plum trees unless considerable lime is added. Both Paris green and London purple are very poisonous, and must be handled with great care.

299*a*. Bordeaux mixture is composed of copper sulfate 6 pounds, quicklime 4 pounds, water 40 to 50 gallons. Dissolve the copper sulfate by putting it in a bag of coarse cloth and hanging this in a vessel holding at least four gallons, so that it is just covered by the water. Use an earthen or wooden vessel. Slake the lime in an equal amount of water. Then mix the two and add enough water to make 40 gallons. It is then ready for immediate use. If the mixture is to be used on peach foliage, it is advisable to add two more pounds of lime to a dilute mixture. Paris green or London purple may be added to the Bordeaux mixture, making a compound which is both fungicide and insecticide.

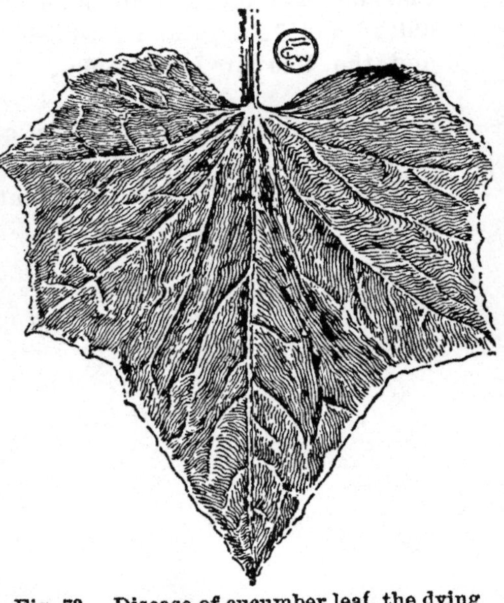

Fig. 73. Disease of cucumber leaf, the dying margin indicating that the trouble is due to some cutting-off of the food supply.

299*b*. Copper carbonate is used as follows: Copper carbonate, 1 ounce; ammonia, enough to dissolve the copper; water, 9 gallons. Before making the solution, make a paste of the copper carbonate by mixing it with a little water. Use 26° ammonia, and dilute with 7 to 8 volumes of water. Then gradually add the necessary amount to the copper carbonate until all is dissolved. It is best made in large bottles. Use only the clear liquid. Dilute as required. For same purposes as the Bordeaux mixture, but does not soil the foliage or fruit.

300*a*. One may find many pictures of pumps in "Principles of Fruit-Growing," Chapter vii., and in Lodeman's "Spraying of

Plants." The latter work should be consulted for the history, principles, and practice of spraying. The bulletins of the various experiment stations may be consulted for the most recent information of insects, diseases, and means of combating them. The reader may also consult various special works, as John B. Smith's "Economic Entomology," Weed's "Insects and Insecticides" and "Fungi and Fungicides," Sempers' "Injurious Insects and the Use of Insecticides," Saunders' "Insects Injurious to Fruits."

303*a*. Smut-infested seeds are treated by corrosive sublimate, formalin, copper sulfate, hot water, and other means. See Swingle, "The Grain Smuts," Farmers' Bulletin No. 75, U. S. Dept. Agric.

CHAPTER XII

PASTURES, MEADOWS, AND FORAGE

I. P. ROBERTS

1. *Grass*

304. The fundamental crop is grass. It covers the land as with a blanket, prepares the soil for other crops, and affords sustenance for farm animals.

305. Grass is one of the important crops in rotations; and a rotation is essential to general husbandry if productiveness of the land is maintained. Rotations improve the farm (*a*) because the land receives different treatments in different years, so that faults of one year may be corrected the following year, (*b*) no one element of plant-food is likely to be exhausted, (*c*) one crop leaves the land in best condition for another, (*d*) roots and stubble of grass, clover and cereals improve the texture of the soil, (*e*) they allow the use of clovers, which add nitrogen, and (*f*) bring up food from the subsoil (170, 170*a*), (*g*) weeds and pests are kept in check, (*h*) labor is economized.

306. The number of plants of grass on a given area should be governed by the uses for which they are grown, their habits of growth and their size. The smaller grasses thrive well if the plants stand near together. The larger grasses, as maize, should have much room between the plants or hills. The plants in a pasture field should be more numerous than in the meadow, and more numerous in the meadow than in fields devoted to raising grass seed.

2. *Permanent Pastures*

2a. *Preparation of the land*

307. When the land is fairly level and can be fitted without too much expense, it is best to plow the ground two or three times during the summer, the first time in early spring, and to keep the surface fine and clean by frequent tillage. This treatment improves the physical condition of the soil, destroys weeds and weed seeds, makes much dormant plant-food available, and conserves moisture so that the surface soil, in most cases, will be damp enough to cause seeds to germinate even in August.

308. On friable soils, as on the western prairies and in some other places, a single plowing and frequent shallow surface tillage may be

the best treatment. On reclaimed boggy lands which have been cultivated long enough to eradicate wild plants, the soil is so light that plowing may be unnecessary. Here a little scarifying of the surface and frequent use of the roller will likely give best results.

309. A good pasture may also be secured by less expensive preparation, if more time is taken. When rolling land has been devoted to the production of cereals and hay until the soil fails to produce satisfactory crops, it is often wise to abandon the unprofitable rotation and to devote the land to permanent pasturage; but few persons are willing to spend as much time and money as will be necessary to secure a good pasture at once. In that case, sow a liberal quantity of pasture seeds in a crop of thinly seeded wheat, rye, barley or buckwheat, the land having been fitted for the cereals with extra care, and plant-food added by a liberal application of fertilizers or manure.

310. Since the pasture is not to be plowed after it is once seeded, it is necessary to prepare the entire soil so perfectly that it will form a comfortable home and provide nourishment for the plants for many years. If the land is poor, fertility should be applied. But prepare the land as best we may, it will not be many years before much of the readily available plant-food

will have been used by the plants, and some of the products of the animals which consume the grass will never be returned to the pasture; hence, the pasture will tend to become less productive as the years pass. And, as the plants become old, they are less vigorous than young ones, not only because of age, but from frequent injuries from the animals. It is, therefore, necessary to maintain the pasture, as well as to prepare it in the beginning.

2b. Maintaining the pasture

311. The grass should be of the right kind. In the North, June-grass or blue-grass is the most permanent pasture grass, and it is the one which gradually works into pastures after other grasses begin to fail. Timothy is commonly sown, about six quarts to the acre. A little June-grass seed may be added, but this grass may usually be depended upon to come in of itself. Orchard-grass is useful in shady pastures and stands grazing well, but grows too much in stools. Red-top is useful in the moister lands. In the South, Bermuda grass and Japan clover are best.

312. After the pasture has been secured, the grasses must be maintained for many years in full vigor. It is pre-supposed that the clovers have been used to a limited extent in the grass-seed mixtures when the pasture was first made,

since the clovers are host plants to the grasses. They start early and protect the later-growing grasses. Most of the clovers live but from one to three years. The clovers, in common with other legumes, contain a large percentage of potential nitrogen (110, 138, 190). The pasture grasses are much benefited by a full supply of nitrogen, but they can secure little, if any, from the air, and hence must supply their needs as best they can from that found in the soil. It will then be understood how eagerly the hungry grasses feed on the decaying short-lived clovers. It will also be understood why clovers are called host plants.

313. The short-lived host plants may be perpetuated, and the grasses kept young and vigorous, by sowing seeds of the clovers and grasses every two or three years in early spring, and scarifying the surface with a sharp-toothed harrow, this to be followed by the roller. The harrowing will not only tear out some of the superannuated grass roots (270) and old plants and cover the seeds, but it will tend to aërate the surface soil and to correct acidity. From time to time, a light dressing of farm manures or of commercial fertilizers should be applied, spread evenly, in the fall.

314. An inspection of the field should be made each spring, in order that seed may be

sown where not enough plants are present, and also to discover what kinds of plants are most promising, so that the supplementary seeds may be chosen to best suit the conditions. Coax the grass to grow by shading the imperfectly covered knolls with refuse material, such as is always found about a farmstead. Even a light covering of brush or maize stalks may be used to partly shade the ground, and to conserve moisture. If a small ration of grain be fed the animals which graze the pasture, the field will tend to become more productive instead of less productive.

315. It will require several years of watchful care, new seed, possibly harrowing and rolling, some added plant-food and a light dressing of lime, and the timely destruction of large, unpalatable weeds, to secure a really good, permanent pasture. The eye of the husbandman makes the grass thrive.

316. In the pastures the grass is kept short; therefore the entire surface should be covered. If areas of even a few square inches are bare, needless evaporation takes place. If the grasses are kept too short, the rays of the sun will take up much soil moisture which should have been taken up by the plants, since the soil will not be well shaded. If the plants are allowed to grow tall and produce seed, then they are

weakened. To prevent the tall growth, mow the pasture, if there are not enough animals to prevent the grass from seeding, and leave the cut material to shade the soil. Aim to preserve the living grass shade intact. Substitute young plants for the old ones. Prevent the soil from becoming acid by light applications of lime and by harrowing it. And, so far as possible, exercise timely care to prevent the plants from becoming hungry and thirsty.

317. Here, then, in a nut-shell, are the elements of a good, permanent pasture: superior preparation of soil, suitable and abundant seeds sown in August, and light pasturing the first season, or, better, mowing the first year; and appropriate seeds and plant-food must be added from time to time, as required.

3. *Meadows*

3*a*. *Temporary meadows*

318. In grain-growing districts, the meadow may occupy from one to three years in a rotation. In dairy districts, meadows are often permanent. The average yield of hay in the North is little more than one ton per acre, although some meadows yield from two to three tons, and, in rare cases, four tons. The average

yield is unprofitable, either in a rotation or in a permanent meadow. As a crop in the rotation, the meadow may improve the soil for subsequent crops.

319. The larger yields are usually secured from vigorous young meadows which contain three or four parts of timothy and one part of mixed clovers. If clover be associated with timothy in approximately these proportions, nearly as much timothy will be secured as if it were sown alone, and the clover, or host plants, will be extra. True, the clovers mature more quickly than the timothy, and this is somewhat objectionable; therefore, the clover mixture may be composed largely of alsike clover, which remains green longer and cures lighter colored than the medium red clover does.

320. The meadow must be viewed from many standpoints. For the city market, unmixed hay sells for more than the mixed, though the latter may be better and more palatable. The uses to which the hay is destined must be considered, since horses should not be fed much clover, while sheep and cattle should not be fed hay composed wholly of timothy and similar grasses. But the meadow remains productive longest where the host plants are present.

321. Whether it is best to leave the meadow for some years and preserve its productiveness

by adding new seed, harrowing, and by the application of plant-food, or to mow it for one or two years and then plow and use the land for other crops, are questions which must be answered by the condition of the meadow and the character of the rotation. There is one invariable rule to be followed,—if the meadow fails to return two tons of field-dried hay to the acre, plow it up; and when the old plants are subdued and the soil put in ideal condition, and when the causes which prevented full success with the old meadow are fully considered, cast in the new seed with understanding, trusting that fuller success will be reached.

3b. Permanent meadows

322. With permanent meadows many new problems are presented. Many fields are of such a character as to preclude a rotation of crops. In such cases the problem is presented of continued liberal production without plowing. Low lands, or those which are wholly or in part overflowed for brief periods, constitute the larger part of our permanent meadows. These low lands are the home of many natural grasses which do not thrive on the uplands; and some of the cultivated upland grasses and the clovers are not at their best when grown in wettish soils.

323. In lowland meadows, a battle royal, which is most interesting and instructive to watch, goes on from year to year. Most of the plants hold their places so tenaciously, and so many hardy new ones appear, that the plants soon become too numerous and then dwarf one another, in which case the production is diminished. On these moist lands there is little difficulty in securing sufficient plants: the problem is rather how to destroy some of them, that better conditions may be secured for those which remain.

324. It has been shown (316) why the pastures should be fully covered with plants; but permanent meadows should have fewer plants. If there are too many, the grasses will not grow to their full size, and many of the leaves on the lower half of the stalks will be yellowish, insipid, and lacking in aroma because they have not received enough sunlight. If there are too many roots in the soil, there will not be sufficient food for all except when the soil is extremely fertile and moist; and few plants will come to normal maturity. The grasses which are grown too thick, and consequently have been excluded from a full supply of sunlight, are poor in quality, like the apples which grow in the shade on the lower branches.

325. All this goes to show how necessary it

may be to destroy some of the grasses in a permanent meadow. By the vigorous use of a sharp-toothed harrow, much may be done to relieve the "hide-bound" and mossy condition, to destroy plants and to aërate the soil (270, 313). A light dressing of lime will materially assist in liberating plant-food and in correcting soil acidity, as in pastures (313).

3c. *Kinds of grasses for meadows*

326. What kind and quantity of seed should be sown, is the question that is asked more frequently than any other, because it is most difficult to answer. In the grass districts of the United States, timothy or "herd's-grass" usually stands first. It is extremely hardy, long lived, is well adapted to grazing, and yet attains good size in the meadow, and when cut at the appropriate time and not over-cured, it makes superior hay. The seeds are not expensive, and can usually be secured without admixture of weed seeds. Timothy, then, in most cases, may form the foundation. Six quarts per acre, more or less, will suffice when used alone, and it may be sown at any time from early spring until fall.

327. We have seen (312, 319) that clover adds to the longevity and productiveness of the pasture or meadow. If the clovers are used, about

the same amount or a little more seed is sown as of timothy, but the plants are likely to be winter-killed if sowing is made after August.

328. There are various secondary and supplementary grasses, such as blue-grass, orchard-grass, red-top, and tall meadow fescue. Some or all of these may be used in limited quantities. Seeds of all these weigh but fourteen pounds to the bushel, are usually sold in the chaff, are not likely to be pure, and are difficult to distribute evenly. In most places, quite as much blue-grass appears as a volunteer as is desirable, but, except in rare cases, it is not a profitable hay grass. Orchard-grass starts early, tends to grow in hummocks, does well in the shade and in close-grazed pastures, but is the worst of all grasses in the lawn, where only fine, recumbent grasses and white clovers are admissible. Red-top is a good pasture grass and lawn grass, and is well adapted to very wet meadows, although it does not make a first-class hay. Tall meadow fescue is one of the most promising recently introduced grasses for both meadow and pasture. In many places it has escaped from the fields into the roadsides, where it shows its superiority over blue-grass and even over timothy. Of these grasses, from one to two bushels of seed are required per acre. All do well when sown in early spring or in fall.

329. Other grasses, as sheep fescue, sweet vernal grass, and similar dwarf grasses, are not to be recommended for general use in America. Other grasses are adapted to special localities, as barley and wild oats, which are extensively used in California for hay. There is a wealth of native grasses, but most of them give little promise for upland meadows.

4. *Other Forage Plants*

330. The plants already discussed, together with other coarser plants of the farm which are fed to domestic animals, are known collectively as forage plants; although this term is commonly applied to such plants as are not grown in permanent meadows or pastures. By recent common consent the term "roughage" has been substituted for them. Both terms are somewhat indefinite. The words usually imply somewhat unconcentrated, dried materials, to which some concentrated food must be added if ample growth, development and surplus products, as milk, are secured.

331. When forage plants are fed green in the stable they are called soiling plants. There are several species of plants, as, for instance, the prickly comfrey, which, if fed green, may

be used for soiling, but, if dried, are unpalatable.

332. The production of forage and soiling crops is extremely simple. They may be inter-tilled or not. Large plants, which require abundant food and moisture and a full supply of sunlight, as maize, should be tilled; but small and quickly maturing ones, as barley, may be raised without inter-tillage.

333. The two great forage plants of the United States are maize and alfalfa. The latter is well suited to the semi-arid districts of the West, and thrives to an astonishing degree in the bright sunshine of the Plains, when supplied with moisture by irrigation. It is perennial, and several cuttings may be taken each season. It is one of the leguminous crops, and, therefore, appropriates nitrogen of the air. Like clover, it has a deep root-system.

334. But the king of all grasses, the one most useful, most easily raised and harvested, and the most productive, is Indian corn, or maize. In a little more than one hundred days from planting, from four to six tons of air-dried stalks and from forty to fifty bushels of grain may be secured from each acre; or from twelve to twenty tons of uncured material may be secured for the silo.

335. Rye, though not a first-class forage or soiling plant, may be sown in the fall, cut when

in head, and followed by a crop of Hungarian grass, which thrives in hot weather; and this in turn may be followed by oats and peas. There will not be time in the North for the oats and peas to mature, but they will remain green through November, and may furnish late fall pasture, or may be left on the ground to serve as a winter cover-crop (115).

SUGGESTIONS ON CHAPTER XII

304*a*. It is impracticable to treat of specific crops in a text-book. Grass and forage are so fundamental to the conception of agriculture, however, that it will be profitable to discuss them, particularly as the cultivation of them illustrates some of the underlying principles of cropping. For advice as to the handling of particular crops, the enquirer must go to books on the special topics.

304*b*. The true grasses constitute the natural family of plants known to botanists as the Gramineæ or grass family; and this family includes all the cereal grains, as wheat, maize, and rice. In its largest sense, therefore, the word grass includes many plants which are not commonly recognized as grasses.

304*c*. The term grass is popularly used to designate the medium sized and smaller members of the grass family, such as orchard-grass, timothy, and blue-grass, and not the larger grasses, as oats, sugar-cane, and bamboo.

304*d*. The clovers are sometimes erroneously called grasses; and "a field of grass" may contain many kinds of plants. There are many kinds of clover. The common red clover is *Trifolium pratense;* the medium red is *T. medium;* the alsike is *T. hybridum*, with rose-tinted flowers; the white or creeping clover, or shamrock, is *T. repens;* the crimson, used for cover-crops, is

T. incarnatum. With the exception of *Trifolium repens*, these are introduced from the Old World. The Japan clover, now much prized in the South, is really not a clover, but belongs to a closely related genus. It is known to botanists as *Lespedeza*

Fig. 74. A carex, or sedge.

Fig. 75. A common sedge, or carex, in flower and when ripe.

striata. It was introduced accidentally into South Carolina about 1849.

304*e*. There are many kinds of grass-like plants. The greater part of these, at least in the North, belong to the closely related Sedge family. Sedges are easily distinguished by 3-ranked leaves and usually by 3-angled stems, with a pith; and the flowers are very unlike grasses. The sedges are generally worthless as forage plants, although some species in the West and South afford acceptable cattle ranges when grass is not to be had. Figs. 74 and 75 show common types of sedges, such as are frequent in swales.

305*a*. In specialty-farming (4*a*), abundance of plant-food and humus material can be added to the soil, and rotations may not be needed; but in general or mixed husbandry some kind of rotation is essential. Read Chapter xv., "Fertility of the Land."

305*b*. The kind of rotation must be determined by the soil and many other factors. A four-year rotation, in which an exacting crop follows a less

Fig. 76. Timothy (*Phleum pratense*) x⅓.

Fig. 77. June-grass or blue-grass (*Poa pratensis*) x⅓.

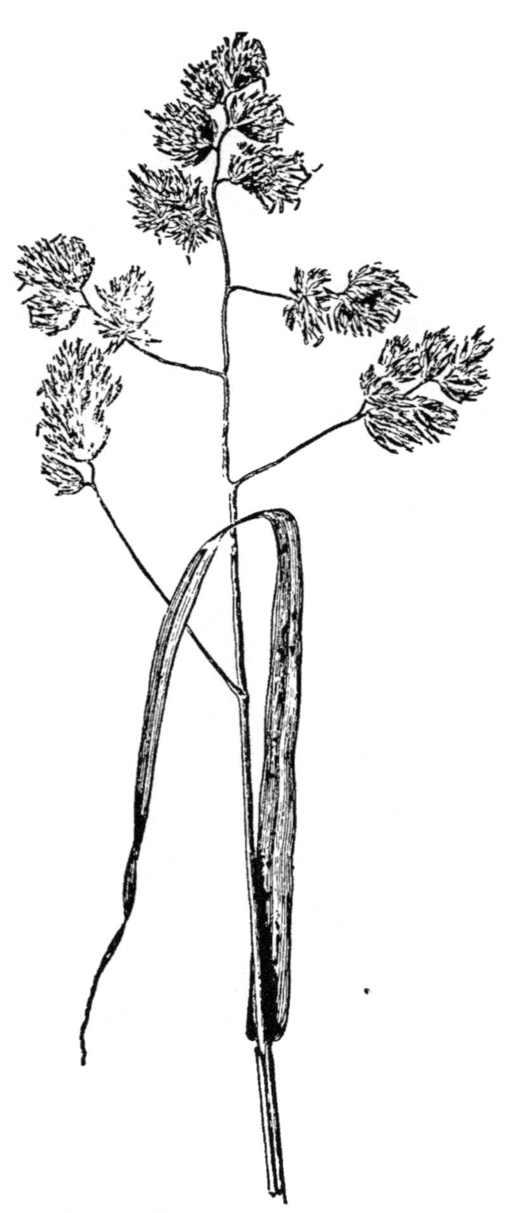

Fig. 78. Orchard-grass (*Dactylis glomerata*) x⅛.

Fig. 79. Tap-root of red clover. (Compare Fig. 33.)

exacting one, and in which the clover root-borer is kept in check, is —

Clover, one year;
Maize, with or without manure;
Oats;
Wheat, with phosphates and manures.

A good rotation for "fairly fertile, lightish lands," is —

Clover, one year;
Potatoes;
Wheat.

A rotation for weed-infested land is —

Sod;
Maize;
Potatoes or some other inter-tilled crop;
Oats or barley.

307*a*. A permanent pasture is one which is to remain many years without plowing. Some pastures, particularly on rocky or rolling land, remain undisturbed for a generation and more. Bermuda grass and Japan clover make permanent pastures in many parts of the South, but most grasses do not make good sod there. In distinction to permanent pastures are the temporary pastures which are a part of a rotation, or the meadow which is pastured after the hay is cut.

311*a*. The familiar Timothy is shown in Figs. 76 and 80. June-grass, with a flower in detail, is seen in Fig. 77. June-grass is a common grass along roadsides, ripening very early, and is the best grass for lawns. Orchard-grass is illustrated by Fig. 78.

312*a*. The word host is here used in a different sense than by the botanist and entomologist (292*b*). Here it means a helper or companion, not a plant upon which another plant or an insect preys.

Fig. 80. Shallow root-system of timothy.

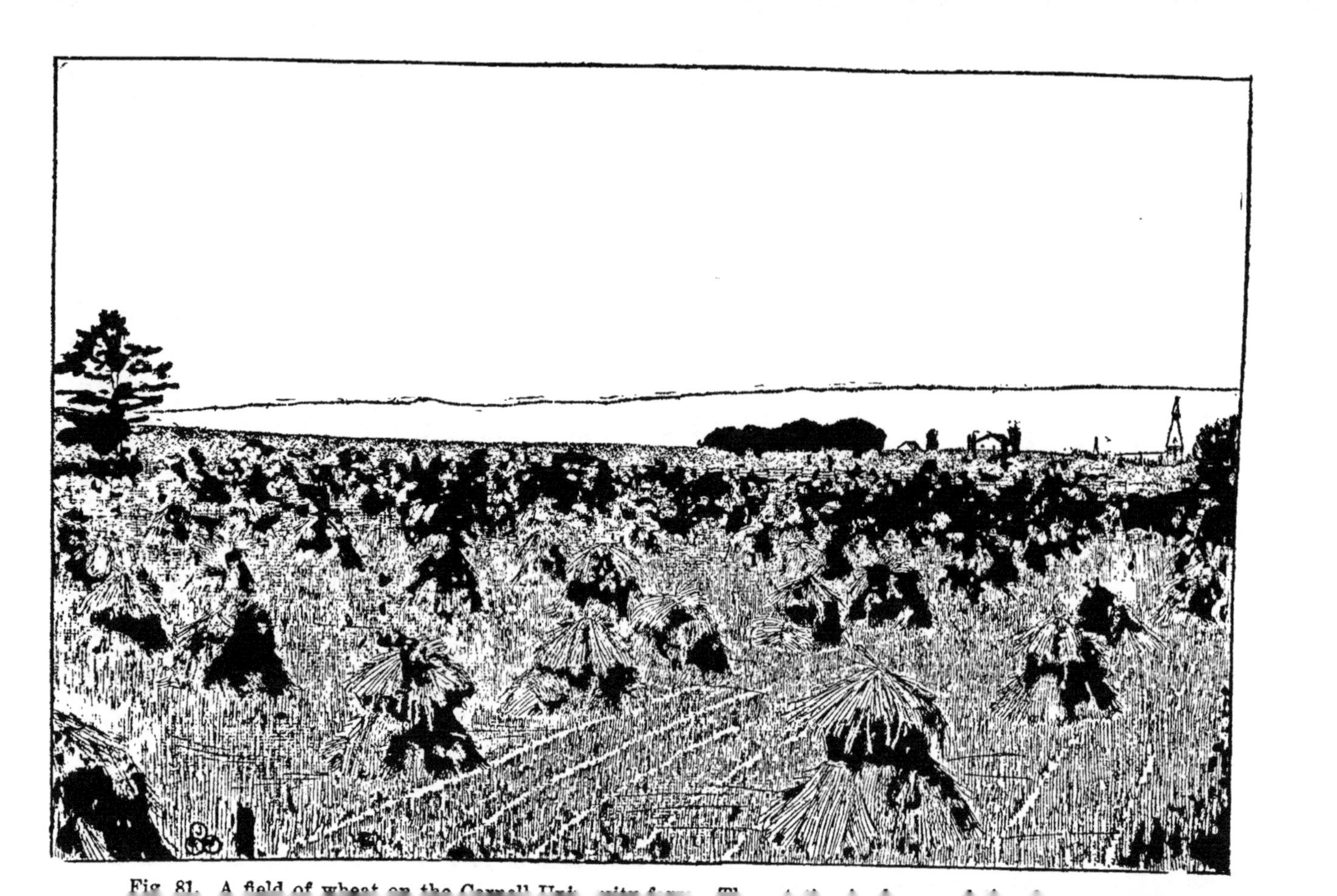

Fig. 81. A field of wheat on the Cornell University farm.

313*a*. Observe how different the roots of clover and timothy are (Figs. 79, 80). One feeds in the subsoil and subsurface soil, has many little organisms on its rootlets, which are called nitrogen-fixers (138); that is, they take the free nitrogen of the air and work it into nitrogenous compounds or albuminoids. The timothy has many small fibrous roots, which remain near the surface, and have no nitrogen-fixing organisms. It will be

Fig. 82. Alfalfa or lucerne (*Medicago sativa*) x½.

Fig. 83. A good bottle for seeds.

seen how appropriate it is to raise these plants together: one feeds near the surface, the other down deep in the soil; one is long lived, the other short lived.

318*a*. In general farming, the most uniformly good crops are nearly always obtained when a rotation is used. Fig. 81 is a field of wheat, in a rotation, which yielded over 30 bushels to the acre.

323*a*. The permanent meadows teach many valuable lessons if they are studied closely. Here is often found a marked illustration of the struggle for existence and of the survival of the fittest. Here the farmer can give little help by tillage, and small opportunity is afforded him to destroy the less desirable plants, that the more desirable ones may have better conditions.

331*a*. It is difficult to keep animals clean when they are fed on green foods; hence the term "soiling."

333*a*. A sprig of alfalfa is shown in Fig. 82. It has small blue flowers in little clusters, and leaves of three leaflets. It is grown somewhat in the East, but it is most useful in the dry regions of the Plains and westward.

335*a*. All the plants mentioned in this chapter should be known to the pupil. In some schools, herbarium specimens may be made of them. It is interesting and useful to collect seeds of farm and garden plants. The school house may very profitably contain a cabinet of seeds. Useful bottles are the "specimen tubes" sold by wholesale druggists and natural-history stores. One is shown in Fig. 83. It is ¾ inch in diameter and 3 inches high, and can be bought, without the corks, for about 30 cents per dozen.

PART III

THE ANIMAL, AND STOCK

CHAPTER XIII

THE OFFICES OF THE ANIMAL

1. *The Animal and the Stock*

336. In an agricultural sense, the animal, as a representative of the animal kingdom, has six general types of uses or offices: it aids in maintaining the fertility of the land; it provides a means of disposing of crops; it, or its products, may be of intrinsic value in supplying food and clothing; it works, or is a "beast of burden"; it may aid in keeping the farm clean of weeds and pests; it diversifies agricultural occupations; it affords employment for labor during the inclement months.

337. When animals are raised in quantity, they are spoken of as stock. This stock may be cattle, turkeys, sheep, ducks, swine, fish, or horses; but in common speech the word is applied mostly to quadrupeds (7).

2. *The Animal in Its Relation to the Soil*

338. The first great resource for the improvement of the texture and richness of the soil is herbage (108–111); the second is farm manures. When stock is pastured, practically all the manure is returned to the farm; but when it is housed, much of the manure is commonly lost through the carelessness of the farmer (120, 120*a*).

339. The greater the proportion of stock to crop, the more fertile the farm should be; for if the farmer must buy feed, the manure is gain, so far as the farm is concerned. In general mixed husbandry, stock is necessary in order to maintain fertility, as well as for its direct value; but in intensive (111*a*) and specialty-farming (4*a*) manures may be bought.

3. *The Animal in Its Relation to the Crop*

340. There is not sufficient market for all the crops which the land can raise. Therefore, some of the crop may be fed to the animal and sold as meat, or butter, or eggs.

341. There is an important secondary gain in this feeding-out of the crop, for part of the crop is returned to the land in the manure. Some

crops, as clover, carry away much more plant-food, if they are sold off the farm, than the animal products which, in large part, are elaborated from them.

4. *The Animal Has Intrinsic Value to Man*

4a. *As articles of food*

342. Animals are direct sources of food. They contribute the various kinds of flesh, as beef, pork, poultry, fish.

343. Animals are indirect sources of food, contributing of their products, as eggs, milk.

344. Animals also contribute materials to various manufactured food products, as cheese, condensed milk, butter.

4b. *As articles used in the arts*

345. Animals contribute materials for clothing. Amongst such products are leather and wool. They also afford material for many articles of personal use, as feathers, bone, hair, glue, horn.

346. Animals contribute largely to fertilizing materials, particularly to substances containing nitrogen and phosphoric acid. Amongst such materials, the most important are bones, dried blood, tankage; of secondary importance are

hair-waste, wool-waste, fish-scrap, hoof-meal, various forms of horn.

4c *As companions*

347. Many animals are pets, or companions to man, and the rearing of them is a species of agriculture. Of such are dogs, cats, rabbits, tame birds, and others.

5. *The Animal as a Beast of Burden*

348. The animal aids in tilling the soil. However much steam may be utilized for propelling implements of tillage, the horse and the ox will still be indispensable to agriculture. Even the tramping of the animals over loose soils tends to compact and improve the land (250*b*).

349. The animal supplies means of transportation. Even with the advent of the electric car, the bicycle and the horseless carriage, the driving horse will remain an important part of the farm equipment.

350. The animal also supplies power for the driving of farm machinery, as threshing and feed-cutting machinery. On large farms, steam power must come to be more and more important, but on the smaller ones animal power will long remain an indispensable factor.

6. *The Animal as a Pest-destroyer*

351. The browsing of animals aids in keeping weeds and wild growths in check. It is well-known that pasturing with sheep is one of the best means of cleaning a weedy area.

352. Animals may keep insect and fungous pests in check by eating the fallen fruit or foliage. It is well known that swine keep the apple-worm in check by eating the windfall apples. Swine also root out and eat the white grub and other insects.

7. *The Animal Diversifies Labor*

353. The animal itself introduces diversity into farming. It also demands the growing of diverse crops. It enforces rotations of crops. Diverse interests educate the farmer, by demanding attention to many problems.

354. Some of the labor which is employed in summer in the growing of crops may be employed in winter in caring for stock. The animal, therefore, introduces continuousness into farming. The best laborers demand employment the year round.

SUGGESTIONS ON CHAPTER XIII

338*a*. It is remarkable how the value of manure increases with the age of the country and the intensity of the agriculture. This comes as a result of experience, wholly without the teachings of science, although science explains why manure is valuable, and points out many of the limitations of its use. The prosperity of the German peasant is measured by the size of his manure-pile. Gardeners place the greatest dependence upon manure; but they want it well rotted,—which means that they not only want its plant-food in the most available condition, but that they desire to utilize it largely for its mechanical effect in loosening the soil with which it is mixed.

341*a*. A ton of clover hay removes about forty pounds of nitrogen, ten pounds of phosphoric acid and forty pounds of potash. A ton of butter removes about two and one-half pounds of nitrogen, and less than one pound each of phosphoric acid and potash.

346*a*. "Tankage is a highly nitrogenous product, and consists chiefly of the dried animal wastes from the large abattoirs and slaughtering establishments. It is variable in its composition, since it includes the otherwise unusable parts of the carcass, as bone, tendons, flesh, hair, etc. The portions of this from the different animals not only vary in their composition, but they are used in varying proportions, which naturally results in an extremely variable product. What is known as 'concentrated tankage,' which is obtained by evaporating the fluids which contain certain extractive animal matter, is the richest in nitrogen, and is more uniform in character than the others; and because of its fineness of division and physical character, the nitrogen contained in it is also more active than in the other forms."—*Voorhees, Fertilizers, 43.*

346*b*. Many other animal substances are used for fertilizers. Those which are used for their nitrogen are dried blood, dried meat, dried and ground fish, sea crabs, hoof meal. Those which are used for phosphates are the various forms and preparations of bone, as raw, boiled, steamed bone, bone ash and bone-black; also, dried fish.

351*a*. With all the remarks which have now been made on weeds (22*b*, 101, 101*a*, 117, 267, 267*a*,268), the pupil will see that the only fundamental and permanent way to escape weeds is through better farm management; and, to a less extent, the same conclusion will apply to insect and fungous pests. "I went by the field of the slothful, and by the vineyard of the man void of understanding; and lo, it was all grown over with thorns, and nettles had covered the face thereof, and the stone wall thereof was broken down."—*Proverbs xxiv., 30, 31.*

354*a*. Upon the desirability of continuous employment for farm labor, Roberts speaks as follows when writing of rotations: "The baleful results of raising a single or few products in extended districts may be seen in California and the great wheat districts of the Northwest. In such localities, there is little or no true home life, with its duties and restraints; men and boys are herded together like cattle, sleep where they may, and subsist as best they can. The work is hard, and from sun to sun for two or three months, when it abruptly ceases, and the workmen are left to find employment as best they may, or adopt the life and habits of the professional tramp. It is difficult to name anything more demoralizing to men, and especially to boys, than this intermittent labor; and the higher the wages paid and the shorter the period of service, the more demoralizing the effect. If there were no other reason for practicing a rotation with a variety of plants, the welfare of the workman and his family should form a sufficient one."—*Fertility of the Land, 369.*

Chapter XIV

HOW THE ANIMAL LIVES

JAMES LAW

1. *The Cell, and Its Part in the Vital Processes*

1a. *The cell*

355. The element in the body that carries on vital processes is the cell; for life in the animal, like life in the plant (Chap. viii.), is dependent on the existence of cells. Each animal cell is a soft, jelly-like substance, held together by an exceedingly delicate network of fibers. It might be compared to a microscopic particle of raw white of egg.

1b. *Single-celled animals*

356. The lowest animals in the scale of existence are formed of a single cell, which in itself performs all the functions of life. This cell can move from place to place, by flowing out from its original globular form, so as to make a projecting arm, and by continuing to flow in the same direction until its whole substance has passed into the new position.

357. This cell can flow out so as to surround microscopic particles and draw them into itself; these it can digest and use to increase its own substance. By reversing this process, it can throw out indigestible and waste materials. It can absorb, digest and build into its own substance nutritive matters already dissolved in water; and it can drive out waste, worn out and injurious matters which it holds in solution in its own liquid.

358. When the cell grows too large, it can divide into two independent parts, each having all the vital powers which belonged to the parent cell or globule.

359. Thus the single-celled animal can make of any part of its body limbs for moving, hands for grasping, fingers for feeling, stomach for digesting, channels for the circulation of its nutritive liquids, as well as organs for excretion and for the increase of its kind.

1*c*. *Many-celled animals*

360. In all the higher animals there is not one cell, but myriads; and these cells are no less essential to life and to the healthy performance of all vital functions than is the single cell of the lowliest organism. In the complex animal body, however, the cells build up solid tissues

outside themselves. As each cell becomes imprisoned in a minute cavity in such solid structure, it is robbed of those common powers or functions which belong to the single-celled animal, and is specialized for the performance of one constant, unchanging round of work. Each cell has its own work to do.

361. Cells may carry on processes of nutrition. Some cells lie in the microscopic spaces left in the hard bone, and conduct the nutrition and changes in its substance. Other cells lie in the substance of muscle or sinew, or of brain, or of some other tissue, and no one of these can construct bone nor any other structure than that in which it lies. All such cells are engaged in carrying on the nutrition and growth of their respective tissues, and are reserved for this work only.

362. Cells may carry on nervous processes, being set apart for vital work of a kind not directly connected with nutrition. Nerve cells,—found in the brain, spinal-marrow, and some other parts,—receive impressions brought over the nerve cords from distant parts of the body. They generate and send out nerve force to other parts. Some of these cells are set in motion by mental acts.

363. Certain other cells, which line microscopic sacs in organs known as glands, select from

the blood the secretion which that gland is appointed to furnish, and pour it out through the gland ducts. The secretion from one gland is nutritious, as in the case of milk; that from another is digestive, as in the secretion of the stomach; and from a third it is waste matter, like sweat. The selection from the nutritive liquid of the blood is the work of the individual cells, and is always the same for each kind of gland.

364. The cells of some glands construct a new substance, which is not secreted but poured back into the blood. Thus the liver makes glycogen, which passes into grape sugar, and serves for the production of heat, muscular work and nutrition.

365. Some cells on the walls of the intestines absorb nutritive and other matters from the liquid contents of the bowels and pass them on into the circulating (blood and lymph) vessels.

366. Besides these cells which become imprisoned in their particular tissues, and the work of which is restricted to the conducting of the growth or other functions of such tissues, there is a large class which floats free in the liquids of the body. The red and white blood globules and lymph cells are examples. These globules or corpuscles circulate in all parts of the body, thus suggesting the freedom of the one-celled animal. But limitations have been

set even to these, the red globules being mainly carriers of oxygen, while the white also have restricted functions.

2. *The Food of Animals*

2a. *Kind of food*

367. Food may be either vegetable or animal. Many animals, as horses, cattle and sheep, live on vegetables, or are herbivorous; while others, like foxes and wolves, eat animal food only, or are carnivorous. The food of the herbivorous animal has its nutritive principles in a less concentrated condition, and the herbivora are accordingly supplied with more capacious digestive organs. The same holds true of grain-feeders and grass-feeders among the herbivora. The grain-fed horse has much smaller stomach and intestines than the grass-fed ox, and the well-fed domestic rabbit has a much more spacious alimentary canal than his wild ancestor.

368. Artificial selection and forcing of meat-producing animals has a similar effect. The scrub ox, Texas steer and buffalo have light abdominal contents, while the pampered short-horn, Hereford, or black-polled ox has them heavy and bulky. In the carnivora they are still more

restricted. The intestine of the ox is about 160 feet long, that of the horse 90 feet, and that of the dog only 12 to 14 feet.

2b. Food constituents

369. All foods must contain chemical constituents which will serve to repair the waste of the body, to develop growing tissue, and to supply materials for the different secretions.

370. Aside from mineral matters, all food constituents which can build up the tissues must contain nitrogen, the element which forms four-fifths of the atmosphere, and which is an essential part of all body tissues. As familiar examples of such nitrogenous foods or aliments may be named white of egg (albumin), milk curd (casein), and one of the soluble parts of flour (gluten).

371. As common forms of foods that contain no nitrogen, and which cannot form tissues, are starch, sugar and fats. These are used up or burned in the system to produce body heat, to stimulate the contraction of muscles, and to furnish secretions which are free from nitrogen, such as sugar and butter-fat in milk, and sugar (more properly glycogen or sugar-former) in the liver.

372. Both sugar and fat, however, can be formed in the body from nitrogenous food, as

in the milk of the carnivorous animal when red flesh only has been fed. In this case the original nitrogenous food is broken up into two or more chemical products, one of which contains only carbon and hydrogen, or these with the addition of oxygen, while all of the nitrogen goes to other product or products.

373. Mineral salts (182*a*) form a third group of food principles. These are essential in repairing the waste of tissues, and in forming secretions like milk, bile and gastric juice.

374. The ideal food contains all of these three groups in forms which can be dissolved, digested and assimilated into the animal tissues. Milk is an ideal food. In it the non-nitrogenous aliments—sugar, butter-fat—are united with the nitrogenous—casein, albumin,—and with the salts in proportions adapted to the needs of the system.

375. A well-balanced ration for the adult animal is one in which these different classes of food constituents bear a somewhat definite relation to each other, due allowance being made for the uses to which the animal is put. The growing, working or milking animal requires more of the nitrogenous elements, while the fattening animal may exchange much of this for the non-nitrogenous.

376. The living body, however, is not like a

simple machine, which can, in all cases, turn out a product exactly corresponding to the chemical food elements which are turned into it. The vital element has always to be reckoned with. One animal demands a little more of this class of aliment, and another a little more of that, in order to secure the best results; while in all cases palatability and facility of digestion have a controlling influence.

3. *Digestion of Food*

3*a*. *What digestion is*

377. Digestion is the process by means of which the food becomes dissolved so as to be taken up by the blood. It takes place in the alimentary canal,—the mouth, stomach, and intestines.

378. Digestion takes place under the action of different secretions, each of which operates on special constituents of the food. Considered in the order in which they mingle with the food, these digestive secretions are: (*a*) saliva; (*b*) gastric juice; (*c*) bile, (*d*) pancreatic juice, (*e*) intestinal juice.

3*b*. *The saliva*

379. Saliva is furnished by a group of glands located under the tongue, in the cheeks, and

under the ears. They discharge their secretions into the mouth. In grain-eating birds, similar glands surround the crop,—an enlargement of the gullet in the region of the neck.

380. A ferment (ptyalin) in the saliva acts on the starch in the food, causing it to chemically unite with additional water and become transformed into sugar. Raw starch is insoluble in water, and cannot pass into the circulation; but the sugar formed from it is freely soluble, can be readily absorbed into the blood, and contributes to the activity, growth and nourishment of the body.

381. The ptyalin acts slowly on raw starch, and much more rapidly on boiled starch, so that cooking of vegetable food favors its digestion. It acts best in the absence of acids. It is less active when weak organic acids are present, and its action is arrested in the stomach by the free muriatic or hydrochloric acid.

382. In animals with one stomach, therefore, it is important that the food should be thoroughly masticated and saturated with saliva, and not bolted whole, or imperfectly insalivated. In ruminants (or cud-chewing animals), as cattle, sheep and goats, the food is long delayed in the first three stomachs, in which any slight sourness which may exist is due to mild organic acids only; and, therefore, there is ample

time and opportunity for the full digestion of the starch.

383. Digestion is further favored in these animals by the chewing of the cud, by means of which the solid portions are returned to the mouth, morsel by morsel, to be leisurely ground down and again saturated with saliva. Digestion is more thoroughly accomplished in the third stomach, in which the food is ground to the finest pulp between the one hundred folds, large and small, which fill its interior.

384. This thorough breaking up or comminution prepares the food for the easy digestion of its nitrogenous principles in the fourth stomach. The removal of the starch renders even the finest particles of food more porous, and permits the prompt and speedy action of the stomach juices on its whole substance.

385. For some time after birth, the salivary glands produce little saliva, and still less ptyalin. This is in keeping with the exclusive milk diet, in which there is no starch to be acted upon. For this reason, any starchy food in the early days of life is out of place; for, as it cannot be changed into sugar, nor absorbed until it has passed through the stomach and reached the intestine, it is liable to ferment and to form irritant products, and indigestion.

The addition of such elements to the foo
should be made later and a little at a time.

3c. The gastric juice

386. The stomach produces three digestiv
principles, which may be separately considered
muriatic or hydrochloric acid, pepsin, the milk
curdling ferment. These materials comprise th
gastric juice.

387. Free muriatic acid is strongly antiseptic
especially checking such fermentations as occu
in the alkaline or neutral saliva, in the first thre
stomachs of ruminants or in the crop of th
bird. This exposure of the food successively t
alkaline saliva and acid gastric juice kills of
myriads of bacterial ferments which would otherwise reach the intestine, to prove irritant o
poisonous. Many still pass into the intestine i
masses of undigested food, or because they ca
survive both alkaline and acid solutions, o
because they have passed into the condition o
spore, which, like the dried seed of plants, i
comparatively indestructible.

388. The muriatic acid further softens, disin-
tegrates, and dissolves the various nitrogenou
food principles (coagulated albumin, fibrin, gela-
tin, casein and vegetable gluten).

389. Pepsin is a ferment which is secreted i

glands found in the end of the stomach nearest to the intestine. It acts on the nitrogenous principles in the food, which are made to take up water, and to change into a much more stable and diffusible liquid called a peptone.

390. Peptones of a great number of different kinds are produced from the varied food principles—from such as fibrin, albumin, gluten, casein. The peptones all agree in certain common characters: (*a*) they are easily and completely soluble in water (fibrin, coagulated albumin and casein themselves, are not soluble); (*b*) they filter rapidly through animal membranes, such as a bladder (the agents from which they are derived do not); (*c*) they are not thrown down as solids by boiling or by strong acids (albumin and casein are precipitated by strong acids, and albumin by boiling).

391. Peptones are thus easily absorbed into the blood, while the absorption of the original principles from which they are derived would be exceedingly slow and difficult. Pepsin acts much more rapidly in an acid medium, so that it is specially adapted to coöperate with the muriatic acid.

392. The milk-curdling ferment is the product of the gastric glands. It is utilized in the manufacture of cheese. Like pepsin, it acts best in the presence of muriatic acid. One part of

this ferment will coagulate 800,000 parts of casein.

393. In birds the gastric juice is secreted in an enlargement of the gullet (proventriculus) just above the gizzard. The strong muscles and cartilaginous lining of the gizzard serve, with the pebbles swallowed, to grind down the food into a fine pulp and to mix it intimately with the gastric juice.

3d. Intestinal digestion

394. Under the action of the saliva and gastric juice, the greater part of the starch and nitrogenous matter is usually digested before the food materials pass from the stomach into the intestines. The products of digestion are mainly sugar and peptones. The fatty matters,—set free by the digestion of their nitrogenous envelopes,—the undigestible portions, and such digestible matters as are as yet not acted on, pass on into the intestines, mostly in a finely divided semi-fluid condition.

395. In the intestines, the materials are acted on by bile, pancreatic juice, and intestinal juice. These fluids are alkaline.

396. Bile is secreted by the liver. It is poured into the intestines a few inches beyond the stomach. It renders the contents alkaline, checks fermentation, stimulates the movements of the

bowels, and transforms their fatty contents into an emulsion which penetrates an animal membrane, and is absorbed with great rapidity.

397. Bile has, besides, a limited power of changing starch into sugar. It is also useful in carrying waste matters out of the body.

398. Pancreatic juice is poured into the intestines by a canal which in certain animals unites with the bile duct. It contains at least four different ferments: (*a*) Amylopsin, which, at the body temperature, rapidly transforms starch and even gum into sugar, thus completing any imperfect work of the saliva; (*b*) trypsin, which, in an alkaline liquid, changes nitrogenous matters into peptones, thus finishing any imperfect work of the stomach; (*c*) a milk-curdling ferment.

399. The pancreatic juice, as a whole, acts like the bile in causing fats to form emulsions. It even breaks up the fats into fatty acids and glycerin.

400. Intestinal juice is a complex mixture of the different secretions already named, together with the products of the glands of the intestinal walls. The secretions of these walls act like pancreatic juice, only less powerfully.

401. As a whole, the digestive agents thrown into the intestines cover the whole field of digestion, and largely make up for any defective

work of the saliva and gastric juice. Even in cases in which the stomach has been removed, the intestines have taken up its functions and have maintained a fair measure of health.

4. *Absorption of the Digested Matters*

4a. *How absorption takes place*

402. The food principles, digested or emulsionized, as before stated, are now absorbed into the blood and lymph vessels, chiefly through the villi of the intestines. These villi are minute hair-like projections from the lining membrane, from $\frac{1}{50}$ to $\frac{1}{30}$ of an inch in length. They are covered with soft cells, the deeper ends of which reach the capillary blood-vessels and lymphatics occupying the interior of each villus.

403. The cells of the villus take in the liquid products of digestion, and pass them on into the vessels beneath. By a muscular contraction of the villus, these vessels are emptied at frequent intervals into the larger veins and lymphatics in the walls of the intestines.

404. The interior of the small intestine, which immediately follows the stomach, is covered throughout by these villi. Owing to the rapid absorption conducted by them, the soluble contents of this intestine are in great

part removed and transferred to the circulatory system before the large intestine is reached.

4b. *Destination of the rich blood from the intestines*

405. The veins from the stomach and intestines carry the rich products of digestion into the capillaries of the liver. Here they not only contribute to produce bile, but also new combinations of nutritive and other compounds, which pass into the general circulation.

406. One of the most important of these new products is sugar, which, as already stated (372), is formed even in the liver of animals fed on a strictly carnivorous diet. The importance of this product may be inferred from the fact that the liver is very large in the young and rapidly-growing animal, and also in mature animals of a meat-producing race: these animals have extraordinary powers of digestion and fattening.

407. Another important function of the liver is the transformation,—largely by union with additional oxygen,—of worn-out or effete red globules, and of much of the useless nitrogenous material in the blood, into urea and other soluble products. These products are finally passed off by the kidneys. They afford a stimulus to secretion by the kidneys, and supply an abundance of material which can pass readily through

these organs without causing irritation or derangement.

408. Another important liver function is the transformation of peptones (which are poisonous when thrown into the blood in any considerable quantity) into products which are non-poisonous, and are capable of assimilation. These products form tissue, or fulfill some other important use in the body.

409. Still another important use of the liver is to transform into harmless compounds the poisonous products of bacterial fermentations (such as ptomaines and toxins). These occur in the contents of the intestine, and might often prove deadly if allowed to pass this guardian sentinel—the liver—in any considerable amount.

5. *Respiration, or Breathing*

5a. *What breathing is*

410. Breathing consists in the substitution of oxygen of the air for carbon dioxid in the blood and tissues of the animal body. It results in the combination of the oxygen of the air with certain organic constituents of the system; and it fits these constituents for various uses, or for elimination as waste matters.

411. In the main, the air is changed in breathing as follows:

	Oxygen	*Nitrogen*	*Carbon dioxid*
Inspired, or breathed-in air contains . . .	20.81	79.15	.04
Expired, or breathed-out air contains . . .	16.033	79.557	4.38

In every 100 parts, air loses, by being breathed, about 4 parts of oxygen, and gains about 4 parts of carbon dioxid.

412. In breathing, the air is also charged with water vapor and with small quantities of ammonia and marsh gas. It also receives a volatile organic matter, which may be fœtid, and when condensed in water soon develops a putrid odor.

413. In the breathing process, the blood and the air are brought into the closest possible contact. One-celled animals breathe through the entire surface, fishes through gills waved in the water, from which they abstract oxygen, frogs through the walls of a simple air-sac, in which the blood-vessels circulate. In warm-blooded animals, this sac or lung is divided throughout into myriads of minute air-sacs or cells, varying from $\frac{1}{200}$ to $\frac{1}{70}$ of an inch in diameter. The walls are so thin that the blood flowing through their capillary vessels is constantly exposed, on two sides, to the air with which they are filled. The membrane constituting the walls of these sacs is so exceedingly

thin and permeable that gases pass through it with great rapidity,—the oxygen from the air to the blood, and the carbon dioxid from the blood to the air.

5b. Blood-changes in respiration

414. The heart of warm-blooded animals is composed of two double cavities, right and left, which are quite distinct from each other. The left side pumps the blood into the arteries of the system at large, whence it returns through the veins to the right side. The right side, in its turn, pumps the blood into the arteries of the lungs, whence it returns by the lung-veins to the left side. In this way the blood is circulated first through the lungs, and then through the tissues of the rest of the body.

415. The blood is of a dark red or purple color as found in the veins, in the right side of the heart, and in the arteries of the lungs. It is of a bright crimson hue as it returns from the lungs and passes through the left side of the heart and the arteries to all parts of the body. The varying color is determined by the presence of a larger amount of oxygen in the arterial (bright crimson) blood, and by its comparative absence, and by the presence of an excess of carbon dioxid, in the venous (dark red) blood.

416. The difference between the artery-blood and vein-blood is shown in the following table:

	Vols. of oxygen	*Vols. of carbon dioxid*
From 100 vols. of arterial blood may be obtained ..	20	39
" " " " venous " " " " ..	8 to 12	46

417. The excess of oxygen in the arterial blood is used up as it passes through the capillaries, and is replaced by carbon dioxid. The excess of carbon dioxid brought back by the venous blood is thrown out into the air filling the lungs, and is replaced in the blood by the oxygen taken up from the air. The carbon dioxid is made up of one atom of carbon obtained by the breaking up of the tissues or blood elements which contain carbon, and of two atoms of oxygen carried to such tissue or element by the blood.

418. Breathing, therefore, or the combination of oxygen with carbon to form the carbon dioxid, really does not take place in the lungs, but in the various parts of the body to which the blood carries the oxygen.

5c. *Amount of air required*

419. The amount of carbon dioxid passed into the blood and exhaled by the lungs is increased by exercise, work, sunshine and food; hence the necessity for more rapid breathing

under such conditions. The amount also varies with the kind of animal. The pig produces more in proportion to his body weight than the carnivora, rabbit, and fowl; and these again produce a larger proportionate amount than the horse or the ox.

420. Air which contains 10 to 12 per cent of carbon dioxid will no longer sustain life. The deleterious effect is due partly to the lack of oxygen in such re-breathed air, but also to the excess of the poisonous carbon dioxid, volatile organic matter, and other injurious products. Air which contains even 1 per cent of carbon dioxid produced by breathing is injurious to a marked degree. In a perfectly close place, where there can be no access of fresh air, a horse would contaminate to this extent over 7,000 cubic feet in 24 hours.

421. The question of stable space, however, is dependent on the amount of air that can be introduced by ventilation in a given length of time. The tighter the building and the less the admission of fresh air, the greater must be the area supplied; while the greater the facility for the entrance of fresh air, the smaller need be the space per animal. If the whole of the air could be removed every three hours, 1,000 cubic feet per horse or cow would suffice to keep the air sufficiently pure and wholesome.

6. *Work; Waste; Rest*

6*a*. *Waste of tissue*

422. Under bodily labor, the elements of the muscles are used up to a certain extent, while heat and waste matters are produced. A period of rest is required to allow for repair of this waste. We see this carried out in all healthy bodily functions. The heart, after each contraction, has a short rest before the commencement of the next contraction. The muscles that carry on breathing work in relays, those that dilate the chest resting while those that compress the chest are in operation. Then both rest for an interval before the next inspiration is commenced. This provides for rest and repair of both the muscles and nerves. Except for such rest, both would soon be exhausted and wasted beyond the power of work.

423. The waste of tissues, however, is not always in exact proportion to the amount of work. On the contrary, it has been shown by careful experiment that the waste of the working muscle is but a small part of the expenditure made. The heat- or fat-producing matters in the food are also used up in such work. The process may be likened to fuel supplied to the engine, which contributes to keep it running

with the expenditure of but a small part of its own proper substance. Thus the starch and sugar in the diet contribute not only to maintain heat and to lay up fat, but also to render possible a large expenditure of muscular energy and work.

6b. Applications to practice

424. Such expenditure of food and muscular energy in producing heat and work prevents the laying out of the same capital for other uses, such as growth, fattening or milking. In domestic animals, which can be profitably kept only when adapted to special uses, expenditures in other directions must be limited as far as may be in keeping with the maintenance of health.

425. For rapid fattening, rest and warmth and seclusion are favorable. Even the milch cow, put in the stable in good health, may be made to give more milk for a time when kept idle in a warm stall than when turned out to gather her food from a pasture. This, however, cannot be safely carried to extremes. The continuous disuse of the muscles tends to their waste and degeneration, to an impoverishment of the blood, to a loss of tone of the nervous and other organs, and to a gradual lowering of vitality. For animals that are soon to be sacrificed to the butcher, this is not to be considered; but for such as

are to reproduce their kind and keep up the future herd, a moderate amount of muscular exercise is as important as suitable food and hygiene.

426. The animal body is a very complex organism, with an almost endless variety of parts and functions, each of which is more or less essential to the full usefulness of the whole. The best condition of bodily health is that in which all of these are properly adjusted to each other and to the surroundings. In the case of farm animals, the complexity is the greater because the natural functions must be developed here and restricted there, to make them a profitable possession; and all this must be done within limits which will be compatible with the maintenance of health and vigor.

SUGGESTIONS ON CHAPTER XIV

359*a*. The best illustration which the pupil can secure of a single-celled structureless organism is the amœba (Fig. 84). This lowly animal lives in stagnant pools, and can be secured by scraping the scum off the stems and leaves of water plants. In its larger forms it is barely visible to the naked eye.

359*b*. The Fig. 85 shows a spindle-shaped (involuntary) contractile cell or fiber from the muscular layer of the intestine, showing nucleus in white and nucleolus in black. It has no such variety of functions as the amœba has.

360*a*. A part or an organism is said to be specialized when it is fitted for some particular work, rather than for general work. A cell which has to do only with nutrition is specialized; one which has to do with nutrition, sensation, locomotion, and reproduction, is generalized. A cell may be said to be

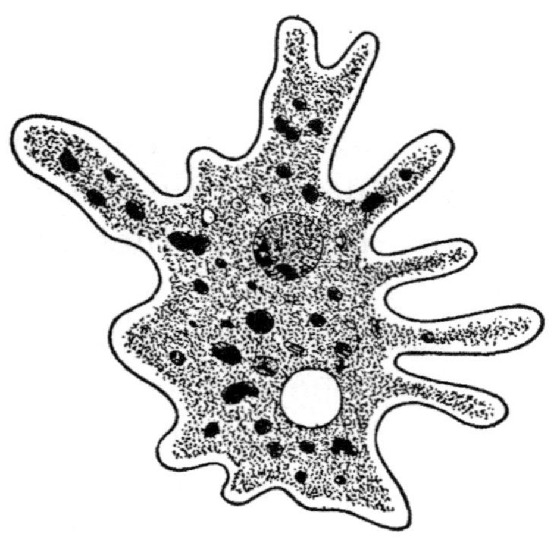

Fig. 84. Amœba, showing large, round nucleus near the top, enclosing a nucleolus, many granules, protruding arms of protoplasm, and white space round which the protoplasm has flowed. Magnified 200 diameters.

Fig. 85. Muscle cell. Magnified.

still further specialized when it carries on some particular or special part of nutrition.

363*a*. A secretion is a material derived from the blood and poured out into the body. When this material is of no further use, it is eliminated, or removed from the body, and is known as an excretion. The saliva, eye-water, bile, gastric juice, are examples of secretions.

363*b*. Glands are secreting organs. Thus the salivary glands secrete or make the saliva or spittle, from the blood. The liver is a gigantic gland, secreting bile and other materials.

364*a*. Glycogen is very like starch. In fact, it has the same chemical composition, $C_6H_{10}O_5$. It is rapidly changed into grape sugar or glucose by the action of saliva and other juices, and it then becomes available for the building of tissue or keeping up the bodily heat.

365*a*. Lymph is a product of the blood. It is a pale liquid which transudes from the thin or capillary blood vessels, and is used to nourish and build up the tissues. The lymphatic system carries food materials to the places where they are needed. See 409*b*.

367*a*. By the alimentary canal is meant the whole digestive tract, beginning with the mouth, and comprising the gullet or esophagus, the stomach, the small and large intestines.

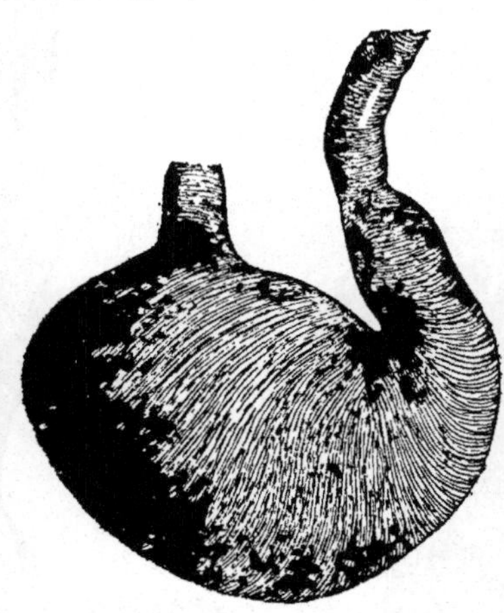

Fig. 86. Stomach of dog.

371*a*. The fats contain carbon, hydrogen and oxygen, but the oxygen is in small proportion. One of the common fats (palmatin) has the composition $C_{51}H_{98}O_6$; another (stearin) is $C_{57}H_{110}O_6$.

379*a*. In physiology, the word ferment is used to designate substances which have power to make starch-like materials soluble by converting them into sugar-like materials. These ferments, of which ptyalin is one, are secretions. They are also called enzyms. These secretions may be the products of cells in the animal body or of independent micro-organisms. The micro-organisms are themselves often called ferments (35*a*).

382*a*. The single stomach of a carnivorous animal is shown in Fig. 86. The stomach of a ruminant is well illustrated in Fig. 87, the front walls being cut away to show the internal structure. It has four divisions: C, paunch ; R, reticulum ; N, manifolds ; O, the true digesting stomach.

385*a*. There are various experiments which the pupil can perform. Mix a little well-boiled starch with a small quantity of saliva, and after a time it will be found to have become sweet. If at the outset a drop of solution of iodine is added to the mixture it will produce a blue color (203*b*). As the starch is changed into sugar, this color gradually fades and in the end disappears.

387*a*. An antiseptic is any material which destroys germs or bacteria (284*a*). The muriatic or hydrochloric acid is present in small amounts, ranging from 0.2 to 0.8 and upward in 1,000 parts in the different kinds of animals.

387*b*. A substance may be acid or sour, in which case it turns blue litmus red (153, 153*a*). It may be alkaline, as lye,

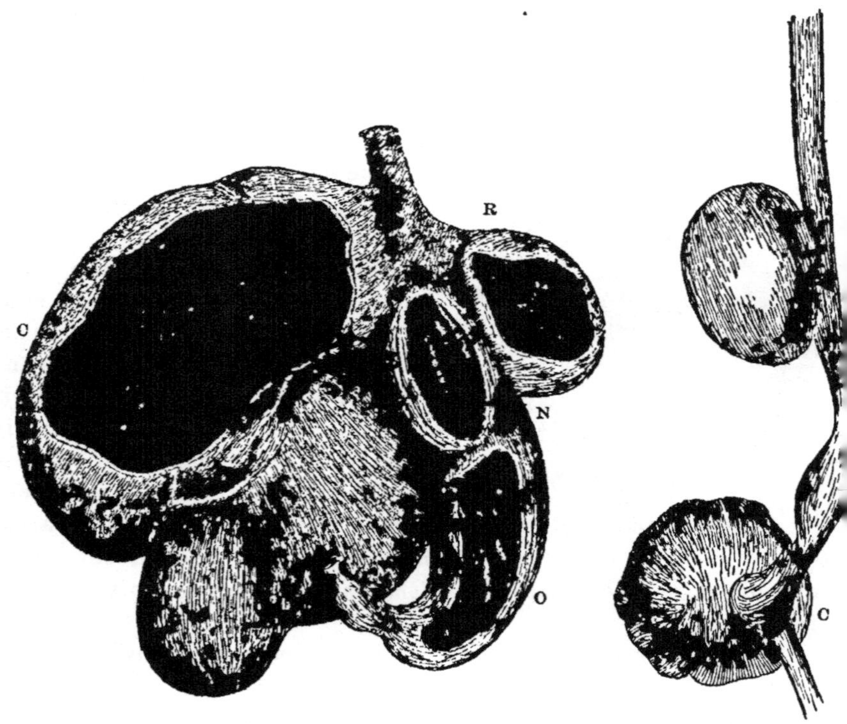

Fig. 87. Stomach of sheep.

Fig. 88. Crop and gizzard of fowl.

in which case it turns red litmus blue. It may be neutral, giving neither reaction.

387*c*. Flowerless plants, of which fungi, ferns, and bacteria are examples, do not produce seeds, but spores. These spores are usually single cells, and contain no embryo. They can usually grow, even after becoming dry. Spores are commonly more difficult to kill than the organism is when in an actively growing condition.

390*a*. A precipitate, in chemistry, is a more or less solid material, which is the result of chemical action, and which settles to the bottom of the liquid in which it is formed. Thus, let the pupil blow through a straw into a bottle of lime water. The liquid will become cloudy, and after a time the sediment will settle to the bottom. The pupil has added the carbon dioxid (CO_2) of his breath to the lime water, and carbonate of lime (or limestone) has been formed. Compare 194*a*.

392*a*. The action of the gastric juice may be familiarly seen in the curdling of milk in the cheese factory by means of rennet. A little mince-meat mixed with the scrapings of the lining membrane of a pig's stomach, rendered slightly acid by a drop or two of muriatic acid and kept near blood-heat (96° F), will soon be completely dissolved, with the formation of peptone.

392*b*. Rennet is the digestive principle derived from the fourth stomach of ruminants (O, Fig. 87). This stomach is taken from calves and dried; and the stomach itself is then spoken of as rennet. The stomach of adult animals could also be used, if necessary.

393*a*. The gastric apparatus of a chicken is shown in Fig. 88. The crop is at *a*, the proventriculus at *b*, and the gizzard at *c*.

396*a*. An emulsion is that condition in which fatty or oily materials are so intimately mixed with the liquid in which they are placed that they act much as if they were in actual solution, even passing through membranes. Most farmers are now familiar with the kerosene emulsion, used as an insecticide (296*a*).

399*a*. Glycerin is a colorless liquid which is associated with fats or fat-acids, and which may be derived from them. Its composition is $C_3H_5(OH)_3$. It is often made from the fats by artificial means, and is used in medicine and the arts. Also spelled glycerine.

402*x*. Two villi are shown in Fig. 89. The singular form of the word is villus.

404*a*. In connection with intestinal digestion and absorption, the bile fills a specially important economic function, in supplying many of its ingredients to be used over and over again in the course of the same day. The bile stimulates in a high degree the absorption of the digested products, entering with them into the veins. As all the blood returning from the intestines must pass through the liver, the elements of the absorbed bile are secreted anew and once more poured into the intestine. Hence a small amount of bile performs a very large amount of work; and hence, too, any suspension of the secretion of bile interferes seriously with the general health.

409*a*. A ptomaine (pronounced tó-main) is a material formed from the decomposition of dead tissue. It is alkaline, and often poisonous. The poison in unwholesome ice-cream, for example, is a ptomaine. Ptomaines often result from the destructive work of microbes. The term toxin is applied to a poisonous product of fermentation, whether alkaline or neutral.

409*b*. It may be well to speak of the destination of the chyle. Chyle is the liquid formed of the materials absorbed from the bowels into the lymph vessels. It is albuminous (nitrogenous) and fatty, with a white, milky color. This, like the lymph in the other lymph vessels in various parts of the body, contains white, spherical, microscopic cells, which are greatly increased after passing through the lymph glands, and when poured into the blood become white blood globules. During the intervals in which there is no digestion, the lymph or chyle in these intestinal vessels, as in other

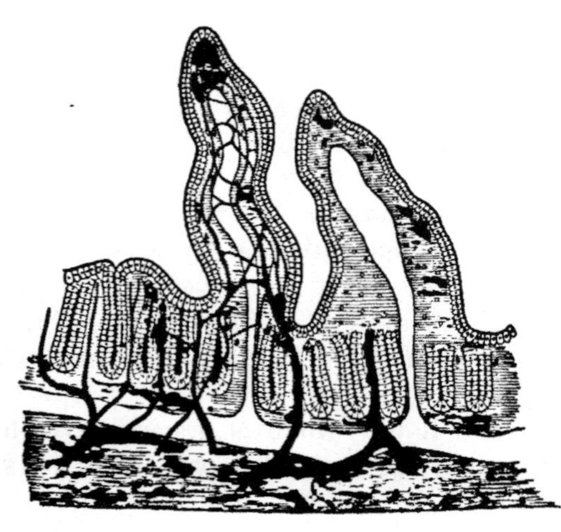

Fig. 89. Surface of mucous membrane of the intestine, showing villi with central lacteal duct and blood vessels, and on the surface the absorbing epithelial cells.

parts of the body, is a simple straw-colored liquid consisting of surplus nutritive matter which has not been required by the needs of the part, and is being returned to the blood. In this lymph we find an important source of supply of the white blood globules, which are being constantly used up; and thus derangements in the lymph vessels and glands injuriously affect the blood, and through it the entire animal system.

409*b*. The admirable adaptation of means to end is traceable in the successive changes of these food products. The nitrogenous constituents in the food, which are not fitted for absorption, are transformed into the peptones, which are specially adapted for rapid absorption. Then the peptones, which are not fitted for nutrition, but are really poisonous, are changed in the liver, so as to render them harmless and fitted for the varied uses of the body, or for elimination. Other food principles are turned into sugar, and some poisonous fermentation products are rendered harmless through the action of the liver. This interdependence of different functions upon each other—mastication, insalivation, digestion, absorption, transformations in the liver, the formation of normal blood elements, assimilation and secretion—furnishes an indication of what goes on throughout the whole animal body, the perfection of one process being essential to that of others, and the derangement of one causing disorder of the others. The nervous system, which is concerned in carrying on all functions, from those of simple nutrition of a tissue or of secretion by a gland up to such mental processes as the animal is endowed with, is dependent on the blood for its own functional activity. Changes in the blood entail change in the capacity for nervous work; so that disorder of one distant organ, acting by influencing the nervous system, directly through the nerves or indirectly through the blood, may bring about derangements of the most varied kind in the different organs subject to nervous influence. The great function of the lungs is the elimination of carbon dioxid from the blood and tissues and the introduction of oxygen, which, being carried into all parts by the red globules, assists in nearly every change which takes place in any organ. But if the lungs fail to fulfill their function to any degree, every organ and function is affected. Most of the waste nitrogenous matter leaves the body through the kidneys, but if this channel of elimination is interfered with, the effete matters are retained, and they poison and derange every organ from the brain downward. Even apparently insignificant organs have a far-reaching influence. The spleen and bone marrow-cells affect the development of blood globules. A small gland at the throat (thyroid) affects the nervous system, and a still smaller one at the base of the brain (pituitary) influences the growth of the limbs.

411*a*. Repeat the experiment suggested in 390*a*. Make lime water by placing a piece of quicklime in a bottle of pure water, shaking and setting aside to settle. Then take a little of the clear liquid and with a syringe force air through it. It will become only slightly turbid. Next take a tube and blow through this water for a short time, when it will become white and opaque by the formation of lime carbonate, owing to the union of carbon dioxid with the lime.

413*a*. The lung of any of the higher animals presents an enormous surface to the inspired air. To illustrate the extraordinary extent of breathing surface formed by this minute division of the lungs into microscopic sacs, it may be stated that, in the horse, it reaches an area of 500 to 800 square feet.

414*a*. The heart of an ox, sheep, or other animal can be obtained at the slaughter house or of the butcher. Discover the right and left cavities,—a ventricle surmounted by an auricle on each side,—the valves around the opening leading from the auricle to the ventricle, and the cords connecting the valves with the inner side of the ventricle.

416*a* When blood is shed in killing an animal or otherwise, observe how the surface layer gradually changes from the dark red to a bright crimson as it takes up the oxygen from the air.

418*a*. In the conveyance of oxygen in the blood the coloring matter of the red globules (hæmoglobin) is the principle bearer. It combines with oxygen loosely, and gives it up promptly at the demand of the carbon. The bright crimson color is due to the union of much oxygen with the coloring matter of the red blood globules, while the dark red hue is caused by the comparative absence of oxygen. The liquid elements of the blood (serum) can absorb and convey but little oxygen. In order to have free and healthy breathing, therefore, the blood must contain abundance of red globules, and these must be well developed, containing a large amount of the red coloring matter. Ill health, lack of sunshine, and various diseases, which cause diminution of the red globules or of their coloring matter, interfere with respiration and consequently with the healthy nutrition and function of the tissues of the animal.

426*a*. Persons who desire a detailed account of the physiology of domestic animals, may consult F. Smith's "Manual of Veterinary Physiology." Advice as to the treatment of animals is contained in Law's "Farmer's Veterinary Adviser."

CHAPTER XV

THE FEEDING OF THE ANIMAL

H. H. WING

1. *Sources of Food of Animals*

427. Broadly speaking, an animal must feed upon either animal or vegetable substances, and it has no power to use as food mineral or inorganic substances.

428. Any substance which an animal may use as food is called a fodder. A fodder must contain the substances that are needed for sustenance in such form that the animal can use them, and must not contain anything that is injurious or poisonous to the animal.

2. *How the Animal Uses Food*

429. The plant, by reason of its vital force and with the aid of the energy of the sun, transforms simple forms of matter into more complex ones, and in so doing locks or stores up a part of the energy received. The animal, by means

of its digestive processes, tears down these more complex substances, setting free the energy, which is turned to its own purposes.

430. Before the energy of the fodder can be used, it is necessary that the animal expend energy upon it during the processes of digestion and assimilation. The profit of the fodder to the animal is represented by the difference between the amount of energy originally present in the fodder and the amount of energy it is necessary for the animal to expend upon it in order to make it available. Some substances require so great an expenditure of energy by the animal to digest or partially digest them that they are useless as fodders, although they may contain the proper compounds in measurably proper proportions.

431. Fodder is used by the animal (1) as fuel to keep up the bodily heat, without which the vital processes cannot go on; (2) to repair the wastes of the various tissues, organs and fluids of the body; (3) to form new tissues or organs, (especially in young animals); (4) to produce young; and (5) to lay up reserve stores in the form of fat or otherwise, to secrete various products, or to perform muscular labor. Some of these reserves or products are useful to man, as milk, wool, and eggs.

432. In general, if the amount of food is

insufficient it will be used for the first four purposes, approximately in the order named; and only after the needs of the animal are fully supplied in these respects will food be used for the last purpose. The food used for the first four purposes is called food of support or food of maintenance; that used for the last purpose is food of production.

433. Not all of the food taken into the body is of use to the animal. The digestive fluids fail to act upon a part of the food, and this passes out through the intestines as undigested solid excrement. It is only the food which is digested that is of use to the animal.

434. The proportion of food digested varies with the animal. One animal may digest 80 per cent of the food eaten; another, standing by its side, equally healthy and equally vigorous and of similar age, may digest less than 40 per cent.

435. The variation in the amount digested may vary with the constituents in the food. Some foods are almost wholly digested; of others less than one-fourth is digested. In any given fodder, one constituent may be readily and largely digestible, while another is digested only with difficulty and in small amount. In general, of the food eaten only from one-half to two-thirds is digested.

3. *Composition of Fodders*

3a. *Classification*

436. Fodders are made up of a large number of substances, all of which are of more or less use to the animal, and each of which, to some extent, serves a definite purpose when used as food. While the number of separate compounds in fodders is very large, they fall into a few very distinct groups or classes, depending upon their composition and the purposes which they serve the animals. These classes are (*a*) water, (*b*) ash, (*c*) albuminoids, (*d*) carbohydrates, including fiber, (*e*) fat.

3b. *Water*

437. Water is present in all fodders without exception, but the proportion is very variable. Some roots and green fresh fodders occasionally have as much as 90 per cent of water, whereas, in some of the kiln-dried by-products the percentage of water may fall as low as 5 or 6 per cent. Ordinary air-dried fodder, as the grains, hay, straw, usually contains from 10 to 15 per cent of water.

438. The water in the fodder to a certain extent supplies the needs of the animal instead of water which is drunk. Animals consuming a

watery food will need to drink less water; but no food contains so much water that it can be used by the animal to supply its needs for both water and solid matters.

439. In general, water adds tenderness, succulence and palatability to fodders. Green fresh fodders are more palatable than the same fodders dried; and the palatability of hay or other dry fodder may be increased by soaking in water, or by steaming.

3c. *Ash*

440. Ash is the small residue which is left when any animal or vegetable matter is completely burned. It is mineral matter obtained by the plant from the soil (147, 192), and is composed of very nearly the same substances in both plants and animals. Some ash is found in all parts of all plants and all animals, and it is necessary to those parts. Life can not be maintained or the vital processes carried on without this ash.

441. In general, the proportion of ash is small, but the bones of animals and certain parts of the plant, as the bark, contain considerable amounts. With scarce an exception, the amount of ash present in ordinary fodders is sufficient for the needs of the animal, and, therefore, it need not be taken into account in

making up a ration or deciding upon a fodder; since no matter what is fed, it is almost certain that the animal will find in it an abundant supply of the proper mineral elements.

3d. Albuminoids

442. The albuminoids, or proteids, constitute a very important group of fodder constituents. While they are of a complex and varied composition, all contain nitrogen as a distinctive constituent, as well as carbon, oxygen and hydrogen, and usually sulfur and phosphorus. It is the nitrogen that gives to the members of this group their importance as food (370).

443. Organic activities can not be maintained without nitrogen. It is an essential constituent of the living animal or vegetable cell, and no new growth can take place without it; consequently it must be constantly supplied in the food of both plant and animal. Nitrogen is not a constituent of the other groups of food elements, and, therefore, the growth of the animal depends in large measure upon the supply of albuminoids in the food.

444. While the albuminoids are found in nearly all fodders, their proportion is very variable, and in very many cases is less than is required by the animal to sustain life or to make useful growth. Those fodders which contain large

amounts of albuminoids are, moreover, much higher priced than those poor in albuminoids. Both these conditions make the problem of successful feeding largely one of the sufficient and economical supply of albuminoids. If an insufficient amount is furnished, the animal suffers in growth or production; if more than enough is supplied, costly waste ensues.

3e. *Carbohydrates*

445. By far the largest part of the dry matter of fodders is classed with the carbohydrates, the most familiar examples of which are sugars, starch, gum and vegetable fiber (371). These substances contain carbon, oxygen, hydrogen—the two latter in the proportions in which they are found in water. They contain no nitrogen.

446. By union with oxygen in the lungs and blood, the carbohydrates are decomposed into carbonic acid (carbon dioxid) and water, and heat is evolved in precisely the same way as under ordinary combustion in the air. They are thus the main source of heat to the animal. They are also the source of muscular energy, and possibly at least a partial source of fat in both tissue and product.

447. Of the carbohydrates, fiber is much less readily acted upon by the digestive fluids, and often a large part of it passes through the animal

without change. For this reason it is often convenient to consider it in a class by itself. So far as it is used at all, it serves the same purpose as the other carbohydrates.

3f. Fats

448. The fats (371*a*) of fodder are used by the animal for much the same purposes as the carbohydrates. They contain only carbon, oxygen and hydrogen, but proportionately much less oxygen than the carbohydrates. For this reason they yield much more energy when decomposed or burned, and are, therefore, of much more value to the animal than the carbohydrates.

449. The amount of energy yielded by different fats varies somewhat, but in general, it is about two and one-fourth times as much as that yielded by an equal weight of sugar or starch; and in reducing fat to its "starch equivalent" (for purposes of comparison) this is the factor commonly employed. In ordinary fodders the percentage of fat is not large, running from about 3 to about 8 per cent of the air-dry substance.

4. *Feeding*

4a. Nutritive ratio

450. From what has already been said, it will be seen that the protein, carbohydrates and fats

are the constituents of the fodder that are of direct use to the animal. These are often collectively spoken of as nutrients, and the portion of them that is digestible as digestible nutrients.

451. Since the protein (or albuminoids) is necessary to growth and reproduction, and since the carbohydrates and fats are mainly used to produce heat and work and reserve stores of fat, the proper relations of these constituents to one another in various fodders and rations constitute an important part of the science and art of feeding. A ration is said to be balanced when these substances exist in the proper proportion to one another for the purpose intended.

452. It has been found convenient to express the relation between the protein and other constituents in the form of a ratio, known as the nutritive ratio. The nutritive ratio is the ratio of the digestible protein to the digestible carbohydrates plus two and one-fourth times (449) the digestible fat, expressed in terms of unity or one of the protein.

453. The nutritive ratio is found by adding to the digestible carbohydrates two and one-fourth times the digestible fat, and dividing by the digestible protein. It is expressed thus: Nutr. Ratio 1: 5.5. It means that in some certain fodder or ration there is for each pound of digestible protein or flesh-forming nutrients, five and

one-half pounds of digestible heat and fat-forming elements. A ratio is said to be wide or narrow when the proportion of heat-forming nutrients is large or small in proportion to the protein. Thus, 1: 12 is wider than 1: 7.

454. A certain proportion should exist between the nitrogenous and non-nitrogenous nutrients of a ration. Animals that are growing rapidly, that are bearing young, and that are producing wool, milk or eggs, require a more nitrogenous food than animals that are working, or fattening, or living without gain or loss of weight. For the latter, the nutritive ratio may be as wide as 1: 12 or 1: 14; for the former, the nutritive ratio should be as narrow as 1: 5 or 1: 6.

455. Formerly it was supposed that slightly differing nutritive ratios would make distinct differences in the effectiveness of a ration or the quality of the product; but it is now generally considered that the limits of variation in the nutritive ratio may be rather wide without materially influencing the nutritive effect of the ration. Other conditions may mask the effect due to differences in the nutritive ratio.

456. One of the chief reasons for taking the nutritive ratio into consideration is that the protein may be economically used. Protein should be used for the formation of nitrogenous products in the animal. It may, however, be used as a

source of heat, instead of the cheaper starch or sugar. This may occur in any ration when the proportion of protein is in excess; but there is generally a too small proportion of protein.

457. By far the larger number of natural fodders are deficient in protein, and a chief task of the feeder is to furnish, from by-products or otherwise, a sufficient amount of albuminoids in the cheapest form. Usually more protein can be used to advantage by the animal than is furnished to it.

4b. Quantity of food required

458. The quantity of food that an animal can profitably or economically use is dependent upon a variety of circumstances and conditions. In the first place, a certain amount must go to the support of the body and the vital functions. This is known as the food of maintenance (432); and a ration calculated to keep an animal alive and in good health without gain or loss of body weight is called a maintenance ration.

459. The amount of food required for support depends upon the size and somewhat upon the individuality of the animal. Small animals require more food in proportion to their weight than large ones. Average animals of the same class, however, are usually considered to require food in proportion to their body weight. In

general, for horses and cattle, about 18 pounds per day of dry matter per 1,000 pounds live weight is required for maintenance.

460. It is from the food eaten in addition to that required for maintenance that the profit comes to the feeder. Hence, if an animal receives no more than enough to sustain life, it can produce no profit to its owner. Much less is there profit if an animal is allowed to lose in weight; for common experience has shown that when an animal is once allowed to suffer loss in weight, the loss is regained only at an increased expenditure of food above what was originally required to produce it.

461. The amount of food that an animal can use profitably over and above that required for maintenance, depends upon the capacity of the animal and the purpose of production. Most animals will make a return approximately in proportion to the food consumed, up to a certain amount. Above that amount, the food simply passes through the animal; or the digestive apparatus becomes disordered and the animal refuses to eat. However, the capacity of different animals in this respect varies widely.

462. Assume that six pounds per day per 1,000 pounds live weight is about the average amount that an animal can profitably use above that required for support. It will be found that

many animals can not profitably use more than three or four pounds, while others can use from ten to fifteen pounds, and an occasional animal can profitably use a still larger amount.

463. The amount of food that an animal can or will eat must not be confounded with the amount of food that an animal can profitably use. Many animals can and constantly do pass through their bodies a considerable amount of food of which no use whatever is made, and this, too, without interfering in any way with the general health, digestive functions, or even with the appetite.

4c. *Feeding standards*

464. Feeding standards show the amount and proportions of the various nutrients that have been found by experience to be best adapted to the various purposes. A few are given:

FOR EACH 1,000 POUNDS LIVE WEIGHT PER DAY.

	Dry matter	*Digestible protein*	*Digestible carbohydrates and fat*	*Nutritive ratio*
Oxen (maintenance)	17.5 lbs.	0.7 lbs.	8.15 lbs.	1:12
Horses at work	22.5 "	1.8 "	11.8 "	1:7
Milk cows	24. "	2.5 "	12.9 "	1:5.4
Growing pigs (young)	42. "	7.5 "	30. "	1:4

465. In any given case, these or any standards may be advantageously varied to a considerable extent. The standards are mere guides.

The skill of the feeder depends upon his success in finding out how far the individual requirements of his animals warrant a variation in the standard.

4d. Bulk in the ration

466. Aside from the amount of digestible nutrients and the nutritive ratio, the bulk of the ration is a matter of considerable importance. It has already been noted (433) that considerable portions of all the nutrients are not digested. Consequently, in every ration there is more or less material of which the animal makes no use, and which may be said to merely add to the bulk of the ration. Water and fiber are, above all other things, the substances which give bulk to a fodder or ration.

467. Fodders which contain large amounts of either or both of these substances are said to be coarse or bulky; fodders which have a minimum amount are said to be concentrated. If a ration is too bulky, the animal is unable to eat enough to obtain sufficient nutrients. On the other hand, a ration may be so concentrated that the proper amount of digestible nutrients do not sufficiently distend the digestive organs so that the gastric fluids may fully act. This is particularly the case with ruminants (382–384, 367).

468. When the ration is unduly bulky because of the presence of large amounts of fiber, it is often so unpalatable as not to be readily eaten. On the other hand, when water is the bulky element, the food is almost always very palatable, but the excess of water has a loosening and depleting effect upon the digestive system. Under ordinary conditions for ruminants, about two-thirds of the dry matter should be furnished in the form of coarse forage and one-third in concentrated food. For horses at work, not more than one-half should be coarse forage, while swine and poultry require the ration to be in a still more concentrated form.

4e. *Palatableness*

469. It is found to be profitable to provide, even at considerable expense, a certain amount of fresh green food for winter feeding, in the form of roots or like material, as a tonic to appetite and digestion. Silage is now popular.

470. The palatability of a fodder or ration,—that is, the readiness or eagerness with which it is eaten,—is a matter of great importance. The nutritive effect of a ration often depends upon this factor alone. In general, animals will make a better return from a ration that is palatable, even though it may not be ideally

perfect according to the standard, than they will from a perfectly balanced ration that they do not like. In many cases the quality of palatability is inherent with the fodder, in others it is due to the individual whim of the animal. It can only be determined for each fodder and each animal by actual trial.

4f. Cooking and preparing the food

471. Most domestic animals are able to eat and digest ordinary forage and grains in their natural state. But almost all fodders may be prepared in various ways so that mastication and digestion are facilitated or palatability increased. Only upon one point is there general agreement—that for most animals it is better that the cereal grains be ground before feeding. As to the advantages and disadvantages of cutting or shredding coarse fodder, and soaking, steaming and cooking foods, opinion is very much divided.

472. There is probably some economy in consumption when coarse fodders are cut or shredded. Palatability is often increased by soaking, steaming or cooking; but cooking renders albuminoids less digestible, and to that extent is a distinct disadvantage.

473. A certain amount of variety in the

constituents of the ration is appreciated by all animals. If the ration is composed of several fodders, these may be mixed in a uniform mass and this mixture fed continuously for long periods of time. This is particularly true of cattle and swine, less so of horses, sheep and poultry.

SUGGESTIONS ON CHAPTER XV

437*a*. By-products are secondary products which result from the manufacture of a given product. Thus, buttermilk and skimmed milk are by-products of butter-making, whey of cheese-making, pomace of cider-making, bran of flour-making.

442*a*. The group takes its name from albumin, which is seen in its purest and most common form in the white of egg. The gluten or sticky part of the wheat kernel, the casein or cheesy part of milk, and the muscular fibers of lean meat, are also familiar examples of albuminoids. From the many forms they assume, they are often spoken of as protein compounds, or proteids. From the characteristic nitrogen, they are also often called nitrogenous substances (370).

443*a*. The albuminoids are necessary to all the processes of growth and reproduction; and since most animal products, as wool, flesh, eggs and milk, contain large amounts of nitrogenous matter, the albuminoids are likewise essential to production as well as growth. When the members of this group are decomposed or broken down, they give up heat, and, therefore, may be used to keep the animal warm (372). It is not at all uncertain that they are not concerned in the formation and storing up of fat in the tissues and milk.

445*a*. The word carbohydrate (written also carbhydrate) means carbon-hydrate. The word hydrate signifies a substance in which water combines with some other element: in the carbohydrates, this other element is carbon. In all the carbohydrates, the oxygen and hydrogen are in the proportions in which they occur in water,—two atoms of hydrogen to one of oxygen (H_2O is water. 130*b*). The carbohydrates are sometimes called amyloids,—that is, starch-like materials.

453*a*. The determination of the nutritive ratio is very simple. For example: clover hay of average quality contains say 7.4% of digestible protein, 11.7% of digestible fiber, 26.3% of digestible carbohydrates other than fiber, and 1.9% of digestible fat. Then 2¼ times 1.9 is 4.3; to this is added 11.7 and 26.3, making in all 42.3, or the starch-equivalent of all the heat- and fat-forming nutrients. Then 42.3 divided by 7.4 equals 5.7. The nutritive ratio of clover hay is, therefore, 1:5.7.

458*a*. The results obtained from any food depend in large measure upon the housing and care which the animal re-

Fig. 90. A cheap and efficient silo.

ceives. Stock should have warm, airy, light, clean, sweet stables (see Fig. 32, p. 86); and in cold weather the drinking water should be slightly warmed. It is cheaper and better to heat the water by means of coal, wood or oil than by means of expensive rations fed to the animal. Stock should not be turned out on cold and blustery days, and a covered yard (Fig. 30) should be provided. To endeavor to secure good results in feeding animals which are cold and uncomfortable is like trying to heat a house with the windows open.

469*a*. Our domestic animals while in a wild state depended for existence almost wholly upon green forage. This trait survives in the fact that in many cases animals will make a larger return for a given amount of nutrients when given green and fresh food than they will for the same nutrients when dry.

469*b*. Silage or ensilage is forage preserved in a green and succulent condition. It is preserved by being kept in a tight receptacle, from which air and germs are excluded as much as possible. This receptacle is called a silo. Maize (corn-fodder) is the most popular silage material. It is cut into lengths of an inch or two and immediately placed in the silo, being firmly tramped and compacted, and the mass then covered with straw, hay, boards, or other material. Circular silos are best because the material settles evenly all around. Fig. 90 shows a very economical silo at Cornell University. It is 12 feet in diameter and 24 feet high, and rests on a cement floor. It is made of lumber 24 feet long, 6 inches wide and 2 inches thick, the edges not bevelled. The pieces are held together by sections of woven fence-wire, drawn together by means of screw clamps. There is no framework. Silage is useful as a part of the daily ration, but it is easy to feed it to excess. Forty pounds a day is usually sufficient for a cow in full milk.

473*a*. Persons who desire to pursue these subjects further should consult Henry's "Feeds and Feeding," and Armsby's "Manual of Cattle Feeding;" also Jordan's "Feeding of Animals."

Chapter XVI

THE MANAGEMENT OF STOCK

I. P. ROBERTS

1. *The Breeding of Stock*

1a. *What is meant by breeding*

474. Animals grow old and die, or they are slaughtered for food. Other animals are born and take their places. Not only is a new animal born, but every pair of animals is able to produce more than two: that is, the total number of animals increases. This birth and multiplication is known as propagation.

475. But it is not enough that new animals and more of them shall appear: these new animals must be desirable. They must have certain attributes or characters which make them valuable. In order that these desirable qualities shall arise, the stockman selects certain animals to propagate the race; and this control of the kind of offspring which shall appear is known as breeding.

476. Breeding may have two objects: to

maintain or reproduce the given type or breed; to produce a new type or breed. One may have small red cows, and desire to produce others like them, or with some improvement on the same lines; or he may wish from these animals to produce large red cows. In the former case, he maintains his type; in the latter, he produces a new type.

477. A breed is a general race or type which reproduces itself more or less closely. It is analagous to a variety in plants. Among cattle, there are such breeds as Short-horns, Jerseys, Devons, Holsteins; among fowls, such as Bantams, Plymouth Rocks, Wyandottes, Shanghais. The person who guides and controls the propagation of animals is known as a breeder.

1*b*. *The mental ideal*

478. The first principle in breeding is to know what qualities one wants to secure. The breeder must have a distinct ideal in mind.

479. Many ideals are impracticable. In order to be practicable or useful, the ideal must be governed by two factors: the person must know the good points of the class of animals with which he is working; he must know which qualities are most likely to be carried over to

the offspring, or be perpetuated. Both of these factors are determined by experience.

480. The ideal type of animal varies with the uses to which the animal is to be put and with the breed. The points of merit in a dairy cow (one which is raised chiefly for the production of milk) are unlike the points in an ideal beef animal. The points in an ideal Short-horn are unlike those in an ideal Ayrshire.

481. Animals are judged by their general form, the texture of the hide and hair, the framework or bony structure, their motions, and their dispositions.

1c. *How to attain the ideal*

482. Having learned what the ideal animal should be, the breeder strives to maintain that ideal by breeding only from those animals which most nearly approach the ideal.

483. Animals vary in their power to transmit their own features to their offspring. Some animals, without any visible cause, possess the power of transmitting their own characteristics to an unusual degree. Such animals are said to be prepotent. Inferior animals may be prepotent, as well as superior ones. It is important, then, to discover beforehand if an animal is prepotent, or is what stockmen call a "good

breeder;" although prepotency can be positively known only by the character of the offspring.

484. In prepotent animals, the eyes are bright, wide open, alert, fairly wide apart and somewhat protruding, or the reverse of sunken. The hair is fine and soft, the skin neither thick and leathery nor too thin or "papery," nor of flabby structure. The bones are of moderate size and have the appearance of being fine grained and strong, as indicated by head, limbs, feet and horns. Such animals are usually symmetrical, although they may not be fat. In all of their movements they are vigorous, alert and powerful and, above all, courageous.

485. Now and then a "sport" appears,—an animal which has some new or strange feature, something which we have rarely or never seen before in that breed (as a hornless or muley animal amongst normally horned animals). Such occasional characters are not likely to be perpetuated. Permanent improvement is secured by slow, small, steady augmentation, not by leaps and bounds.

486. The longer any line of animals is bred to a single ideal or standard, the more uniform the animals become. The breed or the family becomes "fixed." The record of this long line of breeding is known as the pedigree. The longer the pedigree, the greater is the likeli-

hood that the animal will reproduce its characters; that is, characteristics which have been long present are more potent than those which are recently acquired. Hence, a long pedigree, if it records animals many or most of which have been distinguished for some special valuable quality or qualities, indicates more value than a short pedigree.

487. For the general farmer, it is unwise to buy a herd of pure-blood stock, unless the object is to breed pure-blood stock for sale. The breeding of pure-blood animals is a business by itself, and few persons are competent to succeed in it. But every farmer can greatly improve his stock, if he starts with good native animals, by constantly selling off the poorest and breeding from the best. A pure-blood or high-class male or sire placed at the head of the herd will greatly hasten the improvement.

2. *Where Stock-raising Is Advisable*

488. Having now considered some of the principles involved in securing good stock, we may next inquire in what regions and under what conditions it can be raised profitably. Live-stock raising is particularly advantageous on the cheap, unoccupied and uncultivable lands of the West and South. In those regions, stock

must depend largely or entirely on the natural forage, which is sometimes good and sometimes extremely poor and meager. It may require ten to twenty acres to subsist a single cow or steer for a year. If the "range" is eaten off closely during the summer, the animals perish in the winter. In the dry and nearly snowless districts of the West, animals may subsist in the winter on the mature dead grasses. Since the rainfall is light, these matured grasses, or natural hay, retain most of their nutrient qualities.

489. In narrow, sheltered northern valleys surrounded by grass-covered, rolling hillsides, where the cereals cannot be raised to advantage, live-stock finds congenial surroundings. In such regions, for many years, was the center and home of the dairy industries. Within the last twenty years the areas in which butter, cheese and milk have been produced in large quantities for city consumption and export have become greatly enlarged and multiplied; and many whole farms, formerly used for the production of the cereals, especially of maize, are now conducted as dairy farms.

490. On high-priced land near the markets, comparatively little live-stock will be kept, since the manures necessary to keep the soil fairly productive and filled with humus can be easily brought from the cities. The teams which

transport the products to the markets often return loaded with the refuse of the city stables. There is little opportunity for the production of live-stock on the market-garden farm. Where intensive agriculture (111*a*) is carried on, a few animals to consume the refuse, in addition to the "work stock," may be kept to advantage. Swine are often a useful adjunct to market-garden farms.

491. But perhaps the place above all others where live-stock finds the best conditions, and where it is most likely to be improved from generation to generation, is upon the rich, level farms which are adapted to many kinds of crops. Lands which are capable of producing cereals, grasses, fruits, vegetables, flowers and animals should be prized highly. On such lands is offered the greatest opportunity for the highest agriculture. Diversified agriculture, with one or two somewhat specialized crops, leads to steady and certain income, gives opportunity for furnishing continuous employment for both men and teams, and in all ways tends to economy of time and effort (354*a*).

3. *How Much Stock May Be Kept*

492. Cheap transportation, refrigerator cars, and the silo, have made it possible to produce

and send dairy products to market from districts far removed from the great cities and the seaboard, at a profit. On the rich prairies, wherever maize will flourish, one thousand pounds of live stock, or one large dairy cow, may be carried for every two acres of fairly good arable land. In some cases, some extra concentrated foods may be required, if the animals are kept up to their full capacity for growth and production.

493. On farms of the East, where a large percentage of the land must be devoted to permanent pasture because it is steep and stony, one animal of one thousand pounds to two acres cannot be carried unless considerable concentrated food is purchased.

494. There are two theories respecting the number of animals to be kept on a farm. One theory advises that food be bought. The other theory is to keep only so many animals as can be maintained by home resources. On lands naturally fertile, and on those which have been wisely managed, this latter practice is to be commended. It may be said, however, that if the stockman can secure increased profits by risking something for extra food, he should take advantage of it; but most farmers had better not assume many risks.

495. We may now speak of the practice of

purchasing most of the grain or other concentrated food which is required. These foods are mostly by-products (437*a*), such as bran, oil-meal, cotton-seed meal, and the gluten meals. It is said that it is cheaper to purchase concentrated foods than to produce them on the farm, and much stress is laid on the resultant plant-food or manure which is secured from feeding these products.

496. A ton of wheat bran contains the following amounts of potential plant-food in every thousand pounds:

26.7 lbs. nitrogen
28.9 " phosphoric acid
16.1 " potash

This would seem to indicate that a thousand pounds of bran would be worth, for manural purposes, $5.57, or $11.14 per ton—computing the nitrogen at 12 cents, phosphoric acid at 6 cents and the potash at 4 cents per pound.

497. If the bran is fed to milch cows, it is estimated that not less than 50 per cent of the plant-food constituents of the food will be found in the manure. If this be so, then the manure which is the result of feeding one thousand pounds of bran would be worth $2.79, or from feeding a ton of bran, $5.58. If the bran be fed to animals that neither gain nor lose, and are not producing milk or other products, then

nearly all of the manurial constituents of the food are found in the excrements.

498. This practice of purchasing food would appear to be wise on a farm poorly supplied with plant-food. It may be assumed that the increase in growth, or the products secured from the animals which consume these purchased foods, would equal or exceed the cost of such foods. If so, the value of the excrements would be clear additional profit.

499. In practice, however, it is found that the purchase of these supplemental foods becomes necessary largely because a wise use has not been made of the land. If need of these purchased foods arises because but a half crop is secured instead of a full one, then greater attention should be given to making the land more productive. In many cases, the purchased foods are required because the production of grasses and the other forage plants has been neglected. Full crops and wisely purchased concentrated foods lead directly to the improvement of animals and land, and, therefore, to permanent prosperity.

500. When the coarser products are used for food and bedding, and a goodly portion of the grains are fed at home, it is possible, with care, to return to the fields three-fourths of all the plant-food which is removed from the fields

to the barns in the crops. The ease with which a farm may be maintained on a high plane of productiveness when animals are made prominent, and the difficulty of maintaining high productivity when they are wanting, should emphasize the part which the animal plays in securing the best results.

4. *The Care of Stock*

4a. *Housing*

501. Every effort should be exerted to make the animals comfortable. Otherwise, they cannot do their best. Animals,. like people, are most useful when they are happy. Provide them good quarters. As to the style and kind of barns, it matters little so long as the desired results are secured.

502. Animals need much air. A single cow requires in twenty-four hours 3,125 cubic feet; that is, all of the air which would be contained in a box-stall about 18 feet by 17½ feet by 10 feet, if she has a full supply. As a matter of practice, however, a cow is allowed about 400 cubic feet of air. Twice as much air space should be provided in the horse stable as in the cow stable.

503. In the barn, free circulation of air is restricted; therefore, provision should be made for ventilation. Large amounts of air introduced

at few points create dangerous drafts. Air should be taken into and removed from the stable in many small streams. If the stable is over-ventilated, it may become too cold. If at least one cubic foot of air space is allowed in the stable for each pound of live animal kept in it, the air will not have to be changed so often as when the animals are so crowded,—as is often the case,—that only one-half to one-fourth as much air space is provided.

504. A barn with a wall roughly boarded, both inside and outside, and the space filled with straw, furnishes nearly ideal conditions, since the air will be strained gently through the straw. This ventilation should be supplemented by a few small, easily controlled openings. Stables should not be kept above 50 degrees nor fall below 32 degrees, for any considerable length of time.

505. Abundant provision should be made for the ingress of light. It is best if the light is admitted at the rear of the animal, especially for horses. Provision should also be made for temporarily storing the excrements, both to keep the stable clean and to prevent loss of the valuable constituents of the manures. No excrements should be thrown out of the windows or doors of the stable into the open weather, where they form a nuisance and are wasted (120, 120*a*).

4*b*. *Water*

506. All nutriment is carried into the system, and through it, by means of water. Since water is the universal carrier, it should ever be present in the animal tissues in quantities sufficient to accomplish the desired results. Animals should have water at least twice a day.

507. Animals fed a narrow ration (453) require more water than those which are fed a wide ration. A cow in milk may require from fifty to eighty pounds of water daily. The temperature must be raised to between 99 and 102 degrees by the heat generated in the animal. To do this, much of the food of the animal is used which, if warm water had been supplied, would have gone to produce energy in some form, as work, stored fat, or other products. If water raised to blood temperature is provided for the stock in cold weather, the animals will not only enjoy it, but will not require as much food as when compelled to drink water near the freezing point. In large herds, coal may well be substituted for meal in heating the drinking water.

4*c*. *Food*

508. So many varieties of acceptable cattle foods can be secured cheaply in America, that

full opportunity is offered for selecting those which give promise of producing the particular results desired in any given case. Animals which are used continuously at hard work require a wide or carbonaceous ration to supply energy. Young animals do best on a narrow or nitrogenous ration. Milch cows do best on intermediate rations. Cold stables imply a wide ration; warm stables, narrow rations. The food of young herbivorous animals, of those that work, and of cows in milk, may be made up of about one pound of grains or other concentrated foods to three pounds of roughage.

509. The amount of the ration and the time of feeding should be governed according to the character and habits of the animal. Horses should be fed more often than cattle and sheep, since their stomachs are relatively small. Horses are inclined to eat at night. Cattle, sheep and swine seldom eat after dark.

510. The ration for any one meal should not be so liberal as to injure the appetite for the one that follows. Regularity in time of feeding, and skill in presenting the food in an appetizing form, are prime factors of success.

SUGGESTIONS ON CHAPTER XVI

479*a*. The breeder must know the names of the various parts of the animal. The parts of a dairy cow are designated in Fig. 91, which represents a "typical Holstein-Friesian cow:" 1, head; 2, forehead; 3, eyes; 4, face; 5, muzzle; 6, ear; 7, horn; 8, neck; 9, throat; 10, shoulder; 11, shoulder tops, or withers; 12, chest; 13, crops; 14, chine; 15, back; 16, loin; 17, hip or hook; 18, rump; 19, thurl or pin-bone; 20, quarter; 21, thigh; 22, hock; 23, leg; 24, forearm; 25, hoof; 26, fore-

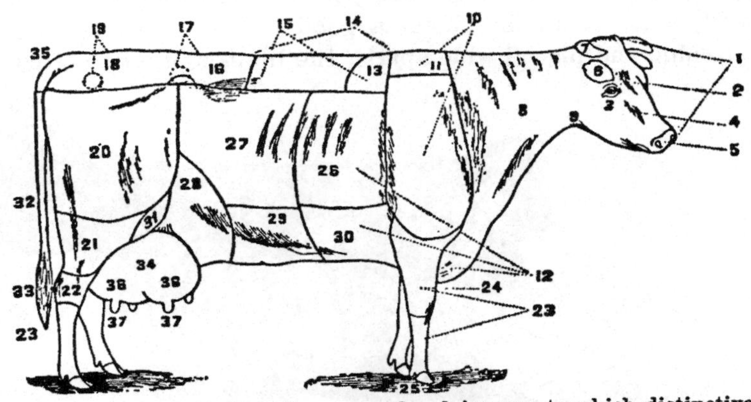

Fig. 91. Diagram to show the parts of a dairy cow to which distinctive names have been given.

ribs; 27, back-ribs; 28, flank; 29, belly; 30, fore-flank; 31, stifle; 32, tail; 33, switch; 34, udder; 35, setting of tail; 36, quarters of udder; 37, teats. The dewlap is the flap of the throat below 9. The escutcheon is the part surrounding the udder behind, on which the hair grows upwards.

480*a*. Following is the ideal of a dairy cow (compare Fig. 92): The cow should have a small head, a large muzzle and mouth, a clean-cut nose or face, that is, one free from fleshy growth, a straight or dishing forehead, bright prominent eyes, and a thin, long neck and moderate-sized horns. She may be from two to four inches lower at the shoulders than at the hips. Her general form, when looked at from the side, should be wedge-shape, and the same shape should be apparent when viewed from the rear. The shoulders may be thin, lean and bony; the back rather long and rugged; the loin fairly broad, but not too broad, or the animal will tend to put on beef. The hip should be thrown well ahead, which gives a long, powerful hind quarter. The thighs, of necessity, are thin; the flank well up; the hind leg, usually, quite crooked, and the tail long. If the tail be long, it is an indication that the vertebræ of the back bone are somewhat loosely united, which is an indication of good milking qualities. The pony-built, smooth-made, short-bodied, rotund cow is seldom a good milker. The teats should be sizeable and placed wide apart; the limbs neither too small nor

Fig. 92. An ideal dairy cow.

too large. The udder should not be very pendent or loose, and should extend well to the rear, also well to the front, and should have a broad and firm setting on the abdomen. The animal should have a rugged, rather lean, but not a delicate appearance. All animals, except those kept for speed, should have rather short limbs, as this indicates, to some extent, constitution and power. It will be noticed (Fig. 92) that the milk veins, which extend from the udder forward on the abdomen, are large and prominent. These indicate that the cow is a great milker or, in other words, that an ample supply of blood has been furnished to the udder by the arteries, and hence a large amount of blood must be returned through the veins. In time, the veins enlarge in order to make room for the return of the blood from the udder. In some of the better milking strains, these large veins are inherited, and can be seen and felt on young animals which have never given milk.

480*b*. Contrast the ideal points of the beef animal. This animal, like the milch animal, should have a small head and horns, and be light in the throat-latch. If the neck, legs and tail be removed from the beef animal, the body is almost a perfect parallelogram. The neck is short and very heavy where it is set onto the shoulder, the back straight, thighs built well out at the rear, and thick. The body of the animal is more rounded, the short ribs or loin is broad, the flank is well down, the shoulders are heavy and well covered with meat, the floor of the chest broad, which places the front legs wide apart. The whole structure of the animal indicates slowness of motion, quietness, and a disposition to lay on flesh and fat, or in other words, to be selfish. No milk veins appear, the tail is shorter than the milch cow's, and the receptacle for milk small. As a rule, the beef animal has a softer and more velvety touch than the dairy animal, since the one is usually fat and the other lean. A strong, low brisket (the hanging part between the fore legs) is desired, not because the flesh of it is good, for it is quite inferior, but because it is an outward indication of superior feeding qualities. It will be noticed that in the dairy cow the brisket is prominent, but thin. It indicates good feeding qualities: that is, a good appetite and power to digest and assimilate food. True, it seems to have no direct connection with the production of milk, but animals which are markedly deficient in brisket and thin in the waist usually have delicate constitutions and precarious appetites.

480*c*. A moderately thick, elastic skin and soft, velvety hair are much desired, not only in cattle but in horses. A thin or papery skin denotes lack of constitution. A thick, inelastic skin denotes unresponsivenessin the production of either milk or beef.

480*d*. With these ideals for cattle, compare some of the points of excellence in a trotting horse: The front legs have a long, low, rhythmic motion when the animal is alert, while the hind quarters are lowered and widened, and the hind legs, with their wide, all-embracing sweep, show how and where the great propelling power is located.

481*a*. The scoring of animals is a matter of ideals. The person assumes that a total of 100 points represents the perfect animal, each part or quality being represented by a certain figure. Any animal may then be judged (as at a fair) by this standard or score. Definite scores have been adopted by various breeders' associations. For illustration, two scores are now given.

481*b*. Following is the score for a dairy cow used by the College of Agriculture, Cornell University:

GENERAL APPEARANCE:	
Weight, estimated ....lbs.; actual............lbs.	
Form, wedge shape as viewed from front, side and top..	5
Quality, hair fine, soft; skin mellow, loose, medium thickness, secretion yellow; bone clean.............	8
Constitution, vigorous, not inclined to beefiness..........	8
HEAD AND NECK:	
Muzzle, clean cut; mouth large; nostrils large..........	1
Eyes, large, bright..................................	1
Face, lean, long; quiet expression......................	1
Forehead, broad, slightly dished........................	1
Ears, medium size; yellow inside, fine texture..........	1
Neck, fine, medium length; throat clean; light dewlap..	2
FORE AND HIND QUARTERS:	
Withers, lean, thin..................................	1
Shoulders, light, oblique	2
Hips, far apart; level between hooks	2
Rump, long, wide....................................	2
Pin-bones or thurls, high, wide apart..................	1
Thighs, thin, long	2
Legs, straight, short; shank fine......................	1
Tail, long, slim; fine switch.........................	1
BODY:	
Chest, deep, low; girth large..........................	8
Ribs, broad, well sprung, long, wide apart; large stomach.	5
Back, lean, straight, chine open	3
Loin, broad, level..................................	2
Flank, moderately low	1
Navel, large...	1

Milk-secreting Organs:	
Udder, long, attached high and full behind, extending far in front and full; quarters even	15
Udder, capacious, flexible, with loose, pliable skin covered with short, fine hair	13
Teats, large, evenly placed	4
Milk veins, large, tortuous, large milk wells	6
Escutcheon, spreading over thighs, extending high and wide; large thigh ovals	2
Total	100

481*c*. The score for a beef steer as used by the Department of Agriculture, University of Wisconsin, is the following:

General Appearance:	
Weight, estimated ....lbs.; according to age	6
Form, straight top-line and under-line; deep, broad, low, set stylish	8
Quality, firm handling; hair fine; pliable skin; dense bone; evenly fleshed	8
Temperament, quiet	5
Head and Neck:	
Muzzle, mouth large; lips thin; nostrils large	1
Eyes, large, clear, placid	1
Face, short; quiet expression	1
Forehead, broad, full	1
Ears, medium size, fine texture	1
Neck, thick, short; throat clean	2
Fore Quarters:	
Shoulder Vein, full	3
Shoulder, covered with flesh, compact on top; snug	4
Brisket, advanced, breast wide	2
Dewlap, skin not too loose and drooping	1
Legs, straight, short; arm full; shank fine, smooth	3
Body:	
Chest, full, deep, wide; girth large; crops full	8
Ribs, long, arched, thickly fleshed	6
Back, broad, straight	6
Loin, thick, broad	5
Flank, full, even with under-line	4

HIND QUARTERS:	
Hips, smoothly covered; distance apart in proportion with other parts	4
Rump, long, even, wide, tail head smooth, not patchy	5
Pin-bones, not prominent, far apart	3
Thighs, full	3
Twist, deep, plump	4
Purse, full, indicating fleshiness	2
Legs, straight, short, shank fine, smooth	3
Total	100

486*a*. A correct, long pedigree is also evidence that no crosses outside of the breed have been made within the time covered by the record. Then the longer the pedigree, the longer the time which has elapsed since the breed was formed. All breeds, as Jerseys, Berkshires and the like, start from mixed-blood animals more or less remote. The term "pure breed" simply means that a breed of animals has been bred so long within the variety that a fair degree of uniformity in all leading characteristics has been secured, and power acquired to transmit the leading qualities with a fair degree of certainty.

487*a*. If the farmer has a dairy, let him resolve to breed from no animal which gives less than 4,000 pounds of milk a year. Animals which give less than this amount are often kept at a loss, and they should be disposed of at once. Every dairyman should also test his milk for richness, by means of the Babcock test. Read Wing's "Milk and Its Products," for instruction on the Babcock milk test, and other matters of dairying.

491*a*. There is a marked tendency for farmers to run too much to one thing, following the lead of some person who has been successful in a particular line. In some localities in the East, especially in the great grape and hop districts, the ill effects of specialized agriculture are often seen. When grapes and hops bring prices which barely pay for picking them,—and this not infrequently occurs,—the farmer becomes discouraged, neglects his plantations, and when prices rise to the point where profits should be received, the yield per acre falls so low by reason of the neglect that no financial recovery is possible. In these districts live stock should play an important part.

491*b*. It is found that wherever the areas of special crops are restricted, and rotation and mixed husbandry are not seriously disturbed, fair profits are realized every year, and the average yields of grapes or hops per acre are much above the average of the large plantations. Specialization is seen to have a marked, deleterious effect on the youth of the districts where it is practiced in a large way, and often on the productivity of the soil as well. The introduction of domestic animals in considerable numbers tends to change all this. Moreover, the elevating effect of coming into immediate contact with animal life, especially on the young, should be understood and prized.

500*a*. A crop of 50 bushels of maize per acre, and the accompanying stalks, contains about 64 pounds of nitrogen, 24 pounds of phosphoric acid and 36 pounds of potash. If, when fed to animals, only one-half of the plant-food removed by the crop is returned, then but 32 pounds of nitrogen, 12 pounds of phosphoric acid, and 18 pounds of potash will be lost from each acre. When clover is in the rotation, it will restore most of this lost nitrogen. The plant precedes the animal. He who has mastered the art of producing plants successfully has learned more than half of agriculture.

500*b*. Animals play such an important part in maintaining the productivity of the land that he who farms without giving them a prominent place should be able to furnish good reasons for so doing.

510*a*. Remember that thoughtful care, solicitude, love for the animal, and timely attention to the many details, play an important part in animal industry. That which is gained by superior breeding, food and comfortable buildings may be partly lost if kindness is wanting. "Speak to the animals as you should to a lady, kindly."

GLOSSARY

(*Numbers refer to Paragraphs.*)

Æsthetic. Appealing to the faculties of taste, as of color, music.

Agriculture. Farming. 1, 1*a*.

Albumin. A nitrogenous organic compound, present in both plants and animals. 370, 442*a*.

Aliment. Food.

Alimentary canal. The digestive channel or tract. 377.

Ameliorate. To improve.

Amenable. Open to, liable to : a loose soil is amenable to the action of air, but a very hard soil is not.

Amendment. A substance which influences the texture rather than the plant-food of the soil. 58.

Annual. A plant which lives only one year. Beans and pigweeds are examples.

Antiseptic. A substance which kills germs or microbes. 284*a*, 387*a*.

Available. Capable of being used ; usable. 43*b*.

Axil. Angle above the junction of a leaf-stalk, flower-stalk, or branch with its parent stem.

Biennial. A plant which lives two years. It usually blooms and seeds the second year. Mulleins and parsnips are examples.

Botany. Knowledge and science of plants. 16.

Breaking down. Said of hard soils when they become mellow and crumbly.

Budding. A kind of grafting, in which the cion or bud is very short, and inserted under the bark or on the wood of the stock (not into the wood).

By-product. A product incidentally resulting from the manufacture of something else. 437*a*, 495.

Callus. The healing tissue on a wound. 234.

Capillary. Hair-like. Said of very thin or fine channels, especially those in which water moves by the force of capillary attraction.

Carbohydrate. An organic or carbon compound, in which hydrogen and oxygen occur in the same proportions as they do in water. Sugar, starch, woody fiber are carbohydrates. 197*a*.

Carbon. A gas, C, existing in small quantities in the atmosphere; also in a solid form in charcoal and the diamond.

Carbon dioxid. A gas, CO_2; carbonic acid gas.

Carnivorous. Feeding on flesh. 174.

Casein. Milk curd, the chief albuminoid of milk. It is the main constituent of cheese. 370.

Catch-crop. A crop grown between plants of a regular crop, in the interval of time between regular crops. 109.

Cereal. A grain belonging to the grass family, as wheat, maize, rice, oats, barley, rye.

Chemistry. That science which treats of composition of matter. 18.

Chlorophyll. The green matter in plants. 198, 198*a*.

Cion. A part of a plant inserted in a plant, with the intention that it shall grow. 236.

Climatology. Knowledge and science of climate. It includes the science of weather (local climate) or meteorology. 19.

Coagulate. To curdle; as of milk.

Coldframe. A glass-covered box or frame which is heated by the sun, and in which plants are grown or kept.

Coming true. Reproducing the variety. 215*a*, 227.

Comminute. To break up, fine, pulverize. 29*a*.

Compost. Rotted organic matter. 34*a*.

Conservation. Saving. 82.

Cover-crop. A catch-crop which is designed to cover the soil in fall, winter and early spring. 109, 116.

Cultivator. An implement which prepares the surface of the ground by turning it or lifting it. The spring-tooth harrow is really a cultivator.

Cutting. A part of a plant inserted in soil or other medium with the intention that it shall grow and make another plant; slip. 231.

Dehorning. Removing the horns from animals. 120*a*.

Dependent. Depending on other means than its own, as on the conditions in which it lives. 182.

Denude. To strip, to make bare, to wash away. 26*b*.

Dormant. Latent, sleeping, not active.

Drought. A very dry spell or season.

Ecology. The science which treats of the inter-relationships of animals and plants, and of their relations to their environments. The study of the habits and modes of life of organisms. The migrations of birds, distribution of plants, nesting habits of bumble-bees, are subjects of ecology. Often spelled œcology. 16*a*.

Element. A substance which is composed of nothing else; an original form of matter. 127*a*.

Emulsion. A more or less permanent and diffusible combination of oils or fats and water. 396, 396*a*.

Energy. Power; force. Every moving, changing or vibrating body or agent expends energy or force; and this force is transferred to some other body or form, for nothing is lost. The energy of sunlight is expressed in heat, light, and other ways. The energy that is required to produce the food is expended as bodily heat, muscular or nervous energy, and in other ways.

Entomology. Science of insects.

Environment. The surroundings of an animal or plant,—the conditions in which it lives. Comprises climate, soil, moisture, altitude, etc. 16*b*.

Erosion. Wearing away; denudation.

Evolution. The doctrine that the present kinds of plants and animals are derived, or evolved, from other previous kinds.

Excretion. A secretion which is of no further use to the animal or plant, and which is thrown off; as sweat. 363*a*.

Extraneous. External; from the outside; foreign to. 54, 59.

Extrinsic. Secondary, external, from the outside. The apple has extrinsic value,—that is, it is valuable as a marketable or money-getting article, aside from its value as nourishment. See intrinsic.

Eye. A bud; a cutting of a single bud. 235.

Farm-practice. The management of the farm; the practical side of farming. It comprises the handling of land, tools, plants, animals. 11.

Farmstead. A farm home or establishment.

Feeding standard. The ideal amount and quality of food for a given purpose. 464.

Fermentation. The process by means of which starch, sugar, casein, and other organic substances are changed or broken down, and new combinations made. It is usually attended with heat and the giving off of gas.

Fertility. Ability of the land to produce plants. 105.

Fiber. Elongated or string-like tissues.

Fibrin. An insoluble but digestible albuminoid. It is present in blood-clots.

Flocculate. To make granular or crumbly. 58*a*.

Fodder. Food for animals. 428.

Forage. Plants which are fed to animals in their natural condition, or when merely dried. 330.

Free water. Standing water, or that moving under the influence of gravitation, as distinguished from that held by capillary attraction. 64, 65, 78.

Function. The particular or appointed action of any organ or part. The function of the eye is vision; that of the heart is distributing the blood; that of the root is taking in food.

Fungicide. A substance which kills fungi. 298.

Furrow. The trench left by the plow. 91*a*. [91, 91*a*.

Furrow-slice. The strip of earth which is turned over by the plow.

Gang-plow. An implement comprising two or more individual plows. Figs. 64, 65.

Geology. The science of the formation of the crust of the earth. 20.

Germ. See micro-organism.

Glacier. A slowly moving field or mass of ice; a frozen stream.

Glands. Secreting organs. 363*b*. [39, 39*a*.

Gluten. The soluble nitrogenous part of flour. 370.

Glycogen. A starch, or starch-like material, formed in the animal body, and from which sugar is formed. 364, 364*a*.

Grafting. The practice of inserting a cion or bud in a plant. 236.

Grazing. Pasturing.

Green-crops. Crops designed to be plowed under for the purpose of improving the soil. 74, 109.

Hard-pan. Hard, retentive subsoil. 94*a*.

Harrow. An implement which pulverizes the surface of the ground without inverting it or lifting it.

Heading-in. Cutting back the tips or ends of branches. 288.

Heavy soils. Soils which are hard, dense, lumpy, or those which are very fertile. Does not refer to weight.

Herbivorous. Feeding on plants. 174.

Horticulture. Arts and sciences pertaining to cultivation of fruits, flowers, vegetables, and ornamental plants. It is part of agriculture. 9, 9*c*.

Host. An animal or plant on which a parasite lives. 292*b*. A plant or animal which makes it possible for another plant or animal to grow alongside of it. 312*a*.

Hotbed. A glass-covered box or frame which is artificially heated (usually by means of fermenting manure), and in which plants are grown.

Humus. Vegetable mold. It may contain the remains of animals.

Husbandry. Farming. 1*a*. [33, 33*a*.

Hygroscopic. Holding moisture as a film on the surface. 64, 67.

Inhibit. To prevent or check. 188.

Inorganic. Matter which has not been elaborated into other compounds by plants or animals. All minerals are inorganic; also, air and water. 25*b*.

Insalivation. Mixing with saliva.

Insecticide. A substance which kills insects. 295.

Internode. In plants, the space between the joints. 205.

Inter-tillage. Tillage between plants. 85, 85*a*.

Intrinsic. Peculiar to, internal, from the inside. The apple has intrinsic value,—that is, it is valuable of itself, to eat, wholly aside from the money it brings. See extrinsic.

Irrigation. The practice of artificially supplying plants with water, especially on a large scale. 63, 63*a*.

Irritable. In plants, responding to external agents, as to wind, sunshine, heat. 183, 208.

Larva (plural *larvæ*). The worm-like stage of insects.

Layer. A part of a plant which is made to take root while still attached to the parent, but which is intended to be severed and to make an independent plant. 229.

Leaching. Passing through, and going off in drainage waters.

Leguminous. Belonging to the Leguminosæ or pea family. 110.

Lichen. A low form of plant-life, allied to algæ and fungi. The plant body is usually grayish or dull-colored and dryish. On tree trunks it is usually called "moss." 29*a*. Fig. 3.

Light soils. Soils which are very loose and open, or which are poor in plant-food. Does not refer to weight.

Marking out. Making lines or marks on the land to facilitate sowing or planting. 103.

Medium. A fundamental or underlying substance: soil is a medium for holding water. An agent: a root is a medium for transporting water. 49.

Microbe. See micro-organism.

Micro-organism. A microscopic organism. It may be either plant or animal; but the term is commonly restricted to bacteria or microbes or germs, which are now classed with plants. 35*a*.

Mineral matter. Earthy matter,—iron, potash, lime, phosphorus, etc.

Moldboard. The curved part of the plow which inverts the furrow-slice. 91.

Mulch. A cover on the soil. 83.

Nitrate. A compound in which NO_3 is combined with a base.

Nitrification. The changing of nitrogen into a nitrate. 137.

Nitrite. A compound in which NO_2 is combined with a base.

Nitrogen. A gas, N, comprising approximately four-fifths of the atmosphere.

Nutrient. Food; aliment.

Nutrition. The process of promoting and sustaining growth and work of animal and plant.

Nutritive ratio. The proportion between the proteids and other constituents in a food. 452.

Optimum temperature. The best temperature for the performance of a certain function. 201, 321.

Organic. Pertaining to organisms,—that is, to animals and plants. Organic matter has been elaborated or compounded of inorganic materials, and exists in nature only as it is made by animals or plants. Flesh, wood, starch, protoplasm, sugar, are examples. The chemist defines organic matter as that which contains carbon in combination with other elements. 25, 25*b*, 32.

Ornithology. Science of birds.

Osmosis. The movement of liquids through membranes. 184, 185.

Oxygen. A gas, O, comprising about one-fifth of the atmosphere.

Palatable. Of good or pleasant taste. 376, 470.

Particles of soil. The ultimate or finest divisions of soil.

Pedigree. A recorded genealogy. 486.

Peptone. A diffusible and soluble compound formed from nitrogenous substances by the action of digestive liquids. 389, 390.

Perennial. A plant which lives three or more years. Rhubarb, apple trees and Canada thistles are examples. [143.

Phosphate. A substance containing or composed of phosphoric acid.

Photosynthesis. Making of organic matter from CO_2 and water in presence of light. 198, 199.

Physical. Pertaining to the body or structure of a thing, as distinguished from its life or its spirit. Pertaining to the action of inorganic forces, as heat, light, electricity, movement of water.

Physiology. The science of life-process or of functioning. It treats of organs, and their work and uses.

Potential. Possible; latent. Said of powers which may be brought into action, but which are now dormant. 42*a*.

Precipitate. The sediment resulting from chemical action. 390*a*.

Prepotent. Said of animals which have the power of perpetuating their own characteristics to a striking degree. 483.

Protoplasm. A very complex and changeable organic nitrogenous compound, present in all living things, and necessary to their existence. It exists in the cells.

Proteid. Albuminoid; protein. 442, 442*a*, 450, 451.

Pruning. Removing part of a plant for the betterment of the remainder. 278.

Ptomaine. A product of decomposition of dead tissue. 409*a*.

Ptyalin. The ferment in saliva. 380.

Puddling. The cementing together of the particles of soils, rendering them hard and stone-like. 81.

Range. A pasture, particularly one of large extent. 488.

Ration. The material fed to an animal.

Rennet. The digestive principle derived from the fourth or true stomach of ruminants ; or the dried stomach itself. 392*b*.

Retentive. Holding, retaining.

Reverted. Said of phosphates which are in the process of becoming insoluble. 145.

Root-cap. The tissue covering the very tip of the growing root. 206.

Root pasturage. The area of soil particles exposed to or amenable to root action. 53*a*, 90.

Rotation. A systematic alternation of crops. 112, 305, 305*a*.

Roughage. Forage. 330.

Sanitation. Looking after the health, especially making the conditions such that disease or injury is prevented.

Sap. The juice or liquid contents of plants. 207*a*.

Saturated. Full of water, so that it cannot hold more.

Scarify. To scratch or to harrow lightly.

Secretion. A special product derived from the blood : as saliva, gastric juice. 363*a*.

Seed-bed. The earth in which seeds are sown. 243*a*.

Seedling. A plant grown from seed, and not changed to another kind by grafting or budding. 241*b*.

Silicious. Sandy.

Slip. A cutting.

Soil. That part of the surface of the earth in which plants grow. 24.

Soiling. Feeding green plants in the stable. 331, 331*a*.

Sport. A variety or form which appears suddenly, or is very unlike the type. 485.

Stock. The plant into which a cion is set. 236. The parentage of any group or line of animals or plants. The animal tenants of a farm ; live-stock.

Stoma, stomate. A breathing-pore. 188, 188*a*.

Subsoil. That part of the soil which lies below the few inches of ameliorated and productive surface soil. It is usually harder, lighter colored, and poorer in plant-food than the surface soil.

Subsoiling. Breaking up the subsoil. 97.

Subsurface. The lower part of the surface soil,—just above the subsoil. 250*a*.

Superanuated. Past its usefulness.

Superphosphate. Sometimes used to designate available phosphates, and sometimes to designate materials which contain phosphate but no potash or nitrogen. 143*a*.

Supersaturated. More than saturated, so that the water drains away.

Supplementary. Secondary; used in addition to something else.

Swine. Hogs, pigs.

Tap-root. A root which runs straight downwards, with no very large branches. Figs. 33, 79.

Texture. Of soils, the physical condition: as loose, tough, open, mellow, hard, baked, puddled. 50.

Tillage. Stirring the soil. 84, 84*a*.

Toxin. A poisonous product of decomposition. 409*a*.

Training. Placing or guiding the branches of a plant. 278.

Transpiration. Passing off of water from plants; evaporation. 187.

Trimming. Removing part of a plant to improve the looks or manageableness of the remainder. 278.

Turbid. Muddy, cloudy.

Under-drainage. Drainage from below. The water is carried through the soil, not carried off on the surface. 57, 68.

Urea. A waste nitrogenous compound which is cast out through the kidneys.

Variation. Modification or change in an animal or plant. The coming in of new forms or types. Departure from the normal type.

Viable. Having life; capable of living or growing. 216.

Vital. Pertaining to life or living things: vital heat is the heat of an animal or plant, as distinguished from the heat of the sun or of a fire.

Weed. A plant which is not wanted.

Watersprout. A strong and usually soft shoot arising from an adventitious or dormant bud,—outside the regular place and order of shoots. 286.

Water-table. That part of the soil marked by the upper limit of the free or standing water. 57, 57*a*.

Zoölogy. Knowledge and science of animals. 17.

INDEX

Printed in Great Britain
by Amazon